Instructor's Manual

Alan D. Magid
Duke University

Charles Kazilek
Arizona State University

Kim Cooper
Midwestern University

Fundamentals of Anatomy & Physiology

EIGHTH EDITION

Frederic H. Martini
University of Hawaii

Judi L. Nath
Lourdes College

PEARSON
Benjamin Cummings

San Francisco Boston New York
Cape Town Hong Kong London Madrid Mexico City
Montreal Munich Paris Singapore Sydney Tokyo Toronto

Executive Editor: Leslie Berriman
Editorial Assistant: Kelly Reed
Managing Editor: Deborah Cogan
Production Supervisor: Mary O'Connell
Senior Manufacturing Buyer: Stacey Weinberger
Supplement Cover Designer: 17th Street Studios
Main Text Cover Design: Riezebos Holzbaur Design Group
Cover Photograph Credit: David Sacks/Getty
Marketing Manager: Gordon Lee

ISBN-10: 0-321-51097-6
ISBN-13: 978-0-321-51097-6

PEARSON
Benjamin Cummings

2 3 4 5 6 7 8 9 10 — TCS — 12 11 10 09
www.aw-bc.com

Table of Contents

Preface

Dear Colleagues,

The instructors who prepare students to work in a health care setting can use all the help they can get in support of this essential mission. Drawing on my four decades of teaching experience, I have prepared this *Instructor's Manual* with the intent of providing both the novice and the seasoned instructor with useful approaches to presenting the material. My approaches assume that your students enter your course with a wide range of backgrounds and abilities but that all possess a strong desire to succeed.

This edition of the *Instructor's Manual* has been extensively revised to correspond with the 8th Edition of Martini/Nath *Fundamentals of Anatomy & Physiology.* Its design follows the text chapter-by-chapter. Each chapter begins with a teaching idea or theme rooted in a contemporary issue in biology, medicine, or society that can be used to pique students' (and instructors') interest in the topic. Background reading outside the text is often suggested to amplify the topic. The introduction is followed by a concise list of Instructional Goals/Learning Outcomes, adopted directly from the textbook.

Learning Outcomes are numbered lists that indicate to students what they should be able **to do** (and not just what they should **know**) after studying the chapter. Educators have begun to shift from "objectives" to more concrete "learning outcomes." The Human Anatomy and Physiology Society (HAPS) has a Curriculum Committee devoted to the ongoing work of developing "learning outcomes" for A&P. This feature keeps pace with this emerging educational trend.

A wide assortment of Teaching Strategies then follows to help you teach each topic. These are organized as:

- Lecture and Lab Ideas
- Analogies
- Demonstrations
- Applications
- Common Student Misconceptions and Problems
- Vocabulary Aids

The 8th Edition of Martini/Nath *Fundamentals of Anatomy & Physiology* incorporates a range of powerful illustration formats. One is "side-by-side" figures, which combine photos of human cadavers with illustrations of the corresponding structures. Students appreciate seeing "real" anatomy, especially since most of us teach without access to cadaver material. An in-depth anatomical resource, bundled with the text, is *Martini's Atlas of the Human Body.* It is filled with gross anatomy images of compelling quality, and also includes an illustrated summary of human embryology. Another illustration format is the "Navigator" figure that presents a "big picture" overview at the beginning of complex discussions that is then repeated with specific details highlighted to break the material into smaller, clearer steps. A related illustration type is used in "step-by-step" figures, which break down complex processes into step-by-step illustrations with clear step headings that coordinate with the author's narrative. Another valuable illustration format is "Macro-to-Micro" figures. These bridge the gap between familiar and unfamiliar structures of the body by illustrating smaller anatomical structures within larger ones.

Pedagogical features of the text include "The A&P Top 100." These boxes offer 100 important concepts and principles over the course of the text and function as organizers of the core information of anatomy and physiology. To help students test their comprehension before moving on to the next topic, "Checkpoints" appear at the end of chapter sections (NOT at the end of the chapter by which time it might be too late!). To engage all students, especially those planning a career in the health sciences, the text is peppered with Clinical Notes, both within the narrative and set off in boxes.

Another valuable supplement provided to students is *Interactive Physiology®* 10-System Suite (IP). This valuable media tool includes animations, interactive quizzes, and concise narrations that cover 10 organ systems. IP-10 is on the student website and also on the CD packaged with the textbook.

The program of available instructor supplements in addition to this *Instructor's Manual* is extensive and powerful:

- **Media Manager 2.1** on DVD organizes all instructor media resources by chapter into one convenient and easy-to-use package. Featured are customizable images of all of the illustrations from the textbook, all of the photos and tables from the textbook, all images from *Martini's Atlas of the Human Body,* PowerPoint Lecture Outlines including text, art, PRS-enabled Active Lecture ("Clicker") Questions, and the Computerized Test Bank on CD. The 3D Anatomy Animations with Quizzes illustrate and test students on muscle and joint actions. The 3D Animations of Origins, Insertions, Actions, and Innervations help students visualize these tough topics.

- A completely revised and carefully checked **Test Bank** provides many thousands of multiple-choice, short answer, and essay questions designed for assessment of factual, conceptual, and critical skills covering all 29 chapters. This resource, in both print and on CD as part of the Media Manager 2.1 package, makes it easy to construct reliable tests and quizzes.

- For use in classrooms lacking computers and data projectors, the supplement program includes **transparencies** with ALL of the art from the text. These figures feature enhanced color and enlarged labels suitable for large lecture halls.

With tools such as these, combined with our love for our students and our enthusiasm for teaching, a fruitful learning experience awaits both student and teacher!

Alan D. Magid, Ph.D.
Duke University Department of Cell Biology
Durham, North Carolina

CHAPTER 1

An Introduction to Anatomy and Physiology

Introduction

Homeostasis, the constancy of the extracellular milieu, is the central unifying concept in physiology. Homeostasis is maintained by *negative feedback* control mechanisms. These two concepts are both abstract and unfamiliar to students and thus often difficult to grasp. The following "thought experiment" should help make both these concepts more concrete. First, make clear that cells are the basic units that make up organisms, and that although their internal workings are complex, individual cells are entirely dependent on the rest of the body to supply their needs. Then, tell the students that cells do best when the fluid that surrounds them provides an approximately constant set of conditions. Tell them that an essential tool of modern biotechnology is the ability to grow cells outside the body under artificially maintained conditions. Ask the class to imagine that they have been hired as biotechnologists to set up a system for growing cells. Guide their thinking by pointing out that cells must be kept in an aqueous environment. Ask the students if the aqueous solution has to have any special properties. Go to the board and develop a list of conditions in the cell surroundings that must be kept relatively constant. They should easily come up with temperature, oxygen, and nutrient levels. You could coax them to include pH, removal of carbon dioxide and other metabolic wastes, osmotic pressure, particular ions, vitamins, etc. Pick a parameter that should be held reasonably constant such as the level of glucose in the culture medium. Since cells consume glucose, it must be replaced as fast as it is consumed. What will be needed to accomplish this constancy of the cell's environment (homeostasis)? First, a probe is required that measures glucose levels (a *sensor* or *receptor*); next a controller that takes the reading from the probe and compares it to the preset desired level (the *set point*); finally, a dispenser that adds glucose to the bath in the amount needed to restore the desired level (the *effector*). Tell them that these components form a *negative feedback loop*. It is designated "negative," not because it is bad, but because this arrangement will remove or *negate* the difference between the set point and the actual level. The term "feedback" is used, because the error is *fed back* from the controller to the effector. To make the point that disease often results from a failure of homeostatic regulation, point to diabetes mellitus in which blood levels of glucose are far too high because some component of the negative feedback loop is faulty.

Instructional Goals/Learning Outcomes

1. Explain the importance of studying anatomy and physiology.
2. Identify basic study skill strategies to use in this course.
3. Define anatomy and physiology, describe the origins of anatomical and physiological terms, and explain the significance of *Terminologia Anatomica (International Anatomical Terminology)*.
4. Explain the relationship between anatomy and physiology, and describe the various specialties of each discipline.
5. Identify the major levels of organization in organisms, from the simplest to the most complex, and identify major components of each organ system.
6. Explain the concept of homeostasis.
7. Describe how negative feedback and positive feedback are involved in homeostatic regulation, and explain the significance of homeostasis.
8. Use anatomical terms to describe body sections, body regions, and relative positions.
9. Identify the major body cavities and their subdivisions, and describe the functions of each.

Teaching Strategies

1. Lecture and Lab Ideas

a. Stress the relationship between anatomy (structure) and physiology (function) by stating, "All specific functions are performed by specific structures." To illustrate this principle at the molecular level, the change of a single amino acid in hemoglobin causes sickle cell disease. Or, the loss of a particular cell type brings on diabetes mellitus. Or the severing of a particular nerve in the arm causes complete loss of sensation and movement in the innervated region. Also, use the structure/function relationship to predict one when given the other. For example, given that the function of mitochondria is to produce ATP, and that liver cells are abundant in mitochondria, it can be concluded that liver cells are very active and so consume much metabolic energy.

b. Sectional anatomy is challenging for students to visualize. It is wise to explicitly acknowledge this difficulty. Start by stating that the process of sectioning reduces a three-dimensional object to a two-dimensional artifact. Have them do several drawing exercises to reinforce the visualization process. Elaborate on Figure 1–11(c), pointing out how it takes thought and imagination to identify familiar organs in this unfamiliar horizontal section. Radiologists struggle with this sort of anatomical problem on a daily basis. Use examples of common items such as pasta shapes (compare spaghetti to elbow macaroni and lasagna noodles). Encourage students to draw the image that would result from various planes of sections. Point out that sections of the human body produce two parts: A sagittal section produces a right and a left piece; a transverse section produces a superior and an inferior piece; a frontal section produces an anterior and a posterior piece. Use these terms to give them practice in hearing and using these directional tags.

c. Body cavities are lined with membranes that secrete a *serous* (watery) fluid. Draw an example on the board, such as the heart in the pericardial cavity or the lung within the pleural cavity. Label and describe the difference between the *parietal* and *visceral* membranes. Illustrate that the parietal membrane lines the cavity that contains the organ, while the visceral membrane covers the surface of the organ (*viscera*, organs within a body cavity), and the cavity is the space between the visceral and parietal membranes. Ask them to guess what function the watery fluid might serve. Help them to understand its function of facilitating movement of the organs by getting to visualize the vigorous motions of a beating heart within the thoracic cavity surrounded by the slowly moving lungs. In addition, be mindful that the concept of "dorsal cavity," commonplace in many earlier A&P textbooks, has been dropped from contemporary anatomical thinking. Only the ventral cavity, derived from the embryonic *coelom*, is a true cavity.

d. When discussing the organization of the text, point out that major section headings are now complete sentences. This makes them easier to understand and to prepare the reader for the information that follows.

e. Encourage the behaviors that lead to success by projecting Tables 1–1 and 1–2 in class and spending some time reviewing them. Seeing them in "black and white" will confer added meaning and substance to these valuable suggestions. The keys here are the "active learner is the successful learner," and "repetition, repetition, repetition" leads to understanding and genuine learning. Acknowledge that the task of mastering A&P is tough but essential to undertaking a career as a health care professional. This knowledge is like the mechanic's specialized tools, the hallmark of an accomplished professional.

f. Review Table 1–4 to show that organ systems combine and cooperate to achieve homeostasis, and to give further exposure to the organ system concept, central to this course and textbook.

2. Analogies

a. As an analogy for homeostasis, imagine a circus performer perched on a board that is balanced on a ball. As the performer balances, he never holds perfectly still; his outstretched arms are continuously moving up and down, and his weight is continuously shifting from one leg to the other. He relies on input from several senses (the sensors) to stay upright (the set point) by using his muscles (the effectors) to remain balanced. That is, he uses his feedback control system to *negate* the balancing errors; thus, this constitutes a negative feedback loop. If his muscles were fatigued and the compensating movements occurred too late, this would work against him. The weakness of the effectors would lead to a *positive* feedback loop: his efforts to keep things in balance lead to the opposite outcome, a worse disturbance, and a fall. Often, this is the course of disease.

b. The analogy used in the text of the thermostatic control of room temperature is an excellent illustration of the component parts or pathway involved in homeostatic regulation. Draw this scheme on the board:

thermometer → thermostat → air conditioner/furnace

sensor → control center → effector

Point out the parallels. For example, note that the thermostat compares the actual temperature to the set point and turns on effectors that add or subtract heat. Ask them what the anatomical and physiological equivalents are. Point out that fever is caused by bacterial toxins acting on the hypothalamus to raise the thermal set point, which we then defend by generating heat. Let them recognize that this concept explains shivering when they have a fever. Ideally, this will trigger an "aha reaction" and reinforce the notion that actively participating in your A&P course will lead to practical benefits in understanding both health and disease!

c. As an analogy to the internal environment for cells (interstitial fluid), have the students imagine fish living in an aquarium. State that fish are living units like cells, complex on the inside, but helpless on the outside. They need certain things in their environment to be held within a satisfactory range by external mechanisms. Review what these factors are and relate them to body function. For example, adding oxygen and removing carbon dioxide; pumping fluids; making nutrients available; keeping osmotic pressure right; holding the pH in range; removing ammonia (nitrogen waste); suppressing bacterial growth; etc. Of course, the keeper of the fish is the sensor, the control center, and the effector, all in one.

3. Common Student Misconceptions/Problems

a. Students need to understand that homeostasis is a dynamic process, one that is constantly fluctuating within normal limits. The literal meaning of homeostasis is "staying the same" but we never are physiologically in a state of "unchanging sameness" unless we're dead! For this reason, it is important that students be aware of "normal range" when learning physiological parameters like pH of blood, RBC count, or heart rate. Only when a controlled variable moves out of range will a compensatory response be made.

b. The mediastinum is not a cavity, but rather a region within the thoracic cavity. The mediastinum contains the pericardial cavity and is flanked by the pleural cavities.

4. Vocabulary Aids

a. Encourage the students to find the anatomical landmarks and regions *on their own bodies* while they study the terms in the text. Relative anatomical directions are more easily mastered if they are studied in pairs. Try this: Give one of a pair of relative anatomic terms in class. Point to it. Say it. Now ask them to say out loud the word it contrasts with (e.g., anterior → posterior) and point to it. Encourage the use of the text in class. Lead them through a few examples, and then recommend that they do them again and again at home until they have mastered these first pieces of anatomical lingo, the first tools in their professional toolbox. Remember, the golden word is *repetition* in learning.

b. The word *parietal* (the wall side) sounds like (but doesn't come from) *per*imeter (outer boundary). Contrast this with *visceral*, related to viscera, the "guts."

c. The usage of "negative feedback" and "positive feedback" by laypersons is very different from the specialized meaning they have for physiologists and control system engineers. This problem is best handled by first reminding students that in anatomy and physiology specialized meanings of words are employed that may differ from everyday usage. Say something like "negative feedback doesn't mean 'your shoes are ugly,'" and "positive feedback doesn't mean, 'I love your hair that way.'"

d. "Serous" sounds like "serum"—the watery component of blood.

e. Point out that the combining form–*stat* in thermo*stat*, is also found in the word homeo*stat*ic. It means "stands" or "stays" fixed or constant.

f. Use the examples of "inferior" and "superior" to make the point that often in anatomy and physiology, common words are used in a specialized sense, different from their commonsense meaning.

CHAPTER 2

The Chemical Level of Organization

Introduction

The chemical (molecular) level of organization is the tiniest that we consider in studies of anatomy and physiology. Atomic or molecular structures and events are unseen in daily life, and yet you need to make it clear to your classes that they are profoundly important. They need to understand that all thought, movement, and memory rest upon ions interacting with protein molecules suspended in a boundless sea of fluid phospholipids, the plasma membrane. The enzymatic breakdown of sugars and fats yields the energy that powers our bodies and enables adding new molecules to the body during growth, development, and repair. The endocrine system is built around chemical signals. Finally, the Human Genome Project represents a worldwide effort costing billions of dollars to determine and understand the exact sequence of nucleotides in the giant DNA molecules (the 23 human chromosomes) that contain the genetic information of a human. Now that this difficult problem has been largely solved, science and medicine are moving forward to literally define the chemical basis of life and to forecast the risk of disease with ever clearer understanding.

Instructional Goals/Learning Outcomes

1. Describe the physiological roles of inorganic compounds.
2. Discuss the structures and functions of carbohydrates.
3. Discuss the structures and functions of lipids.
4. Discuss the structures and functions of proteins.
5. Discuss the structures and functions of nucleic acids.
6. Discuss the structures and functions of high-energy compounds.
7. Explain the relationship between chemicals and cells.

Teaching Strategies

1. Lecture and Lab Ideas

a. Direct the students to the periodic table (Appendix III). Using the table as a guide, apply the core atomic concepts that are defined in the textbook: atomic number, chemical symbol, element, atomic weight. In terms of electron orbitals, explain the horizontal and vertical arrangement of the elements, and point out the periodicity. In support of the crucial role that careful observation plays, you can point out that Mendeleev, the author of the best-accepted periodic table in the 1860s, lacked a comprehensive structural model of the atom, commonplace today, but still saw and ordered the periodicities. Learning more about this brilliant man from western Siberia and his pioneering work—never honored with a Nobel prize because he died too soon—would make a good independent or group research paper in a course that uses research and writing assignments.

b. Even though a single hydrogen bond possesses only about 1/100 of the bonding energy of a covalent bond, the total energy of all the hydrogen bonds within a single molecule or between molecules can be quite significant. H bonds maintain the 3-D shape of large molecules and supramolecular ensembles such as proteins and nucleic acids. It is the H bonds that are disrupted when pH drops or temperature rises during denaturation reactions.

c. It is important to illustrate that H bonds occur between separate molecules (or distant regions of a large molecule) whose atoms are joined by covalent bonds. A board drawing of several water molecules, with both the polar covalent bonds and the H bonds clearly labeled, will emphasize this point.

d. Table 2–2 is a biggie for the beginning student. Make it crystal clear that *mastery* of these six simple rules of chemical notation will become essential tools in their professional "toolkit," needed now and down the road in their future training and practice. Too much of medical care has a chemical component to not be completely comfortable with chemical notation, most of which is commonsensical and easily understood, if somewhat unfamiliar.

e. It is hard to overemphasize the significance of enzymes. In the first case, it is the first form of molecular recognition they will encounter in the course. The lock-and-key fit between enzyme and substrate molecule is what confers *specificity* on the catalyzed reaction. Later examples include hormone receptors, channels, carriers, pumps, 2nd messengers, cell adhesion molecules, and neurotransmission. Enzymes are what makes life possible, exploiting a chemical *trick* to accelerate chemical reactions at mild temperature. Living things would not survive in the chemist's boiling cauldron! Enzymes lower the energy hump without changing the initial or final energy status of the reactants and products.

f. When presenting the different classes of reactions, anticipate their roles in metabolism (catabolic vs. anabolic). Also, link each with an important physiological example, such as dissociation/association applied to the chemistry of carbonic acid dissociation/association. This connects to buffering reactions as well.

2. Analogies

a. Anthropomorphize molecules by describing them as both "greedy" and "lazy." They want their outer energy level complete, and will look around for someone else's excess electrons. However, they are lazy; they won't take on more than half the number of electrons required to fill the level. In those cases, they would just as soon give up the electrons in the outer energy level but will be left with one less energy level. A few molecules are lucky enough to come equipped with completed outer energy levels, and so can be standoffish. The noble gases do not interact with anyone!

Likewise, if atoms can't agree on who is going to give up an electron to make an ionic bond, they may form even stronger *covalent bonds* by simply sharing the electron in their outer shells. Some atoms express *electronegativity* when in a compound. This means that bonding electrons spend more time in their neighborhood, conferring a partial negative on them. This induces a corresponding *electropositivity* on other atoms within a compound. "Hydrogen bonds" form between the party-charged oxygen atom and a hydrogen atom of two neighboring water molecules and confer many of water's unusual properties.

b. Compare the primary structure of a protein to a toy train made up of many types of cars (amino acids). When hooking the cars together, there is a definite front and back, just like the C- and N-terminus of a polypeptide chain. The chain of cars may be made as long as you wish and of very different form and function by assembling boxcars, passenger cars, club cars, baggage cars, tank cars, and so on. The hydrolytic digestive enzymes can be analogized to railroad workers that separate the cars at their couplings, thus making the cars available to form new trains.

c. Enzymes will work at their optimum rate as long as there is plenty of substrate, and other conditions are favorable. Think of a roofer fixing a roof. He pounds in nails to hold down shingles (combining two substrates to produce a product), and can use the same hammer again and again (enzymes are catalysts not consumed in the reaction). This process can continue as long as there are nails and shingles.

d. Compare one strand of DNA to a spiral staircase: The alternating sugar and phosphate groups make up the helical supports, while the bases are analogous to the steps. Of course, DNA is made of two antiparallel helices, which would be a very confusing stairway!

e. Enzymes can be thought to work like helpers getting a car from down in one valley, over the hill, and down into a lower valley. The enzyme "lowers the hill," making it easier to get from one side to the other with much less kinetic (heat) energy. Enzymes don't change where you start or where you finish but just make it easier (and thus faster) to get over the hill. Refer to Figure 2–7 during this analogy.

3. Demonstrations

a. Compare the different levels of protein structure with ribbon used to wrap a package. A ribbon can be stretched out straight (primary), coiled in a spiral (secondary), crimped (secondary), or stripped with a scissors so that it develops many overlapping curls (tertiary). You can even bunch several of the curly ribbons together to make a wreath (quaternary).

b. To make dramatically clear and concrete the difficult concept of the mole as an Avogadro's number of particles (and to introduce the value of the "gedanken experiment" in scientific reasoning), try this simple but interesting lecture or lab exercise. Ask the students how they could count out 6.02×10^{23} water molecules. Answer: by weighing exactly 18 grams of water! Spell it out:

1 mole of oxygen (16 grams) + 2 moles of hydrogen (2 grams) = 18 grams

Ask them to visualize how much water this is (a bit over a half ounce). Repeat for other molecules, such as the *monosaccharides* fructose or glucose (180 grams). How big a pile would this make (about 120 mL)? How much for the *disaccharide* sucrose? (342 grams, about 207 mL). Why not 360 (180×2)? Because sucrose is formed in a condensation reaction (a loss of 1 water molecule, -18). How much would a mole of hemoglobin weigh? Answer: 64 kg (141 lb)! A mighty big pile of red stuff. Ask them to imagine that. Very different size piles of molecules but yet all have the same number of particles, an Avogadro's number of them, 6.02×10^{23}. This thought exercise should impress on them the physical meaning of "mole" (and on the power and importance of *imagination* in scientific thinking).

4. Applications

a. During the discussion of free radicals, mention the strong evidence that eating plentiful amounts of fruits and vegetables high in antioxidants (and thus, presumably better able to quench free radicals) protect against cancer, heart disease, and possibly other grave illnesses. Oddly, little benefit, and sometimes harm, results when the antioxidant nutrients are consumed in pure form, away from the foods in which they naturally occur in goodly amount.

b. Indicate that prostaglandins are found in most cells. Stress that their importance as chemical signals will become clear later in the text.

c. Explain the role of cholesterol, both as a membrane component and as starting material for steroid hormones.

d. Explain the structural and functional roles of phospholipids and glycolipids. You can use Figure 2–17(c) to anticipate the phospholipid bilayer by asking the students to imagine what would result if a spherical micelle were flattened.

e. Solutions are very important in biology and medicine. Be sure they are clear about how a solution forms, which is the solute, and which the solvent. The cell cytoplasm and the blood plasma are good examples of complex biological solutions.

5. Common Student Misconceptions/Problems

a. Each integer change in pH value represents a tenfold change in the concentration of hydrogen ions. The *lower* the pH value, the *higher* the concentration of hydrogen ions. If the blood pH is 7.3, although an alkaline pH, it is slightly more acidic than the normal pH of blood (7.35–7.45). Hydrogen ions affect the pH only if they are in solution. A buffer acts like an ion sponge, binding or releasing hydrogen ions, and so limits changes in pH if hydrogen ions are added or removed from solution.

b. Anticipate the possible confusion between the *alpha helix* in the secondary structure of proteins and the *double helix* of the two complementary antiparallel strands in DNA.

c. Clarify that "cholesterol" is not merely a "bad" thing but an important body chemical. It is a component of many cell membranes, plasma lipoproteins, and the source of steroid hormones. It is particular cholesterol compounds that promote cardiovascular disease.

6. Vocabulary Aids

a. For mnemonics for cation vs. anion (that is, positive ion vs. negative ion), let the "**t**" in cation remind you of a "+": ca+ion. Also, let **AN**ion remind you of **A N**egative ion.

8. Describe the stages of the cell life cycle, including mitosis, interphase, and cytokinesis, and explain their significance.
9. Discuss the regulation of the cell life cycle.
10. Discuss the relationship between cell division and cancer.
11. Define differentiation, and explain its importance.

Teaching Strategies

1. Lecture and Lab Ideas

a. Osmosis refers to movement of water across a membrane. Osmosis occurs across membranes that are not freely permeable to solutes, but are permeable to water because of the presence of *aquaporins*, transmembrane protein channels through which water molecules move single file. Water molecules flow toward solutions of higher solute concentration, since such solutions willy-nilly have a lower water activity, that is, higher [more negative] osmotic pressure. In other words, osmotic pressure, due to *dissolved solute*, is a *negative* pressure, which "sucks" the water across the membrane.

   Don't be shy about quickly reviewing simple facts about solutions. Often, it is hard to underestimate the scientific readiness of the average A&P student. Even the quite bright ones, especially in courses that require no prerequisites in general chemistry and cell biology, can be quite unfamiliar with basic chemistry and cell science. This and the previous chapter constitute "mini-textbooks" of these two fields. Use them as intensively as your circumstances suggest. At the least, mention that they are available for self-study and make specific assignments to master the key figures and tables even if not covered in lecture. Some teachers may want to use the lab to introduce some of the concepts. It's a natural way to introduce weights and measures (temperature, heat, energy, pH, mass, length, solutions), and cells (RBCs, cheek cells) and microscopy.

b. Vesicular transport involves "bulk" transport as compared to the molecular movement seen in diffusion, active transport, facilitated diffusion, osmosis, and so on. Thus, by forming a vesicle enclosing a volume of cytoplasm or extracellular fluid, the membrane itself becomes a part of the transport machinery in contrast with solutes that move across the membrane particle-by-particle through channels or by interacting with carrier proteins.

c. Functions of microfilaments include:

   Anchor the cytoskeleton to integral membrane proteins
   Interact with other proteins to produce cellular movement

   Functions of intermediate filaments include:

   Provide strength
   Stabilize position of other organelles

   Functions of microtubules include:

   Provide cell strength, rigidity; anchor organelles
   Alter shape of cell; assist in cell movement
   Align and move organelles within the cell
   Form the mitotic apparatus for cell division
   Form centrioles, cilia, and flagella

d. The function of microvilli is to increase cell surface area; they are located on cells that are involved in transporting substances across the cellular membrane from extracellular fluid. Microvilli contain bundles of microfilaments.

e. Functions of the endoplasmic reticulum include:

   Synthesis of proteins, carbohydrates, and lipids
   Storage of materials
   Transport of materials within the cell
   Detoxification of drugs or toxins

   Functions specific to smooth ER include:

   Synthesis of phospholipids and cholesterol
   Synthesis of steroid hormones, in gonads and suprarenals
   Synthesis and storage of triglycerides, in liver and fat cells
   Detoxification of drugs, in liver and kidney cells
   Synthesis and storage of glycogen, in muscle and liver cells

   Functions specific to rough ER include:

   Synthesis and modification of proteins
   Packaging proteins for transport to the Golgi apparatus

   Functions of the Golgi apparatus include:

   Synthesis and packaging of secretions for exocytosis
   Renewal or modification of the plasma membrane
   Packaging of enzymes for use in the cytosol

f. Eukaryotic DNA (including human) is always bound to histone proteins.

   Nucleosomes are formed by DNA wrapping around histone clusters.
   In interphase cells, each strand of DNA, with its histone proteins, has a characteristic fine filamentous appearance, and is called extended chromatin or *euchromatin*.

   Nucleoli are composed of histones, RNA, and enzymes, coiled around the DNA regions that code for ribosomal proteins and ribosomal RNA.
   The chromatin *supercoils* only during mitosis (or in the mature sperm nucleus), giving the chromosomes their characteristic condensed shapes.

g. To illustrate the details of protein synthesis, as well as provide a brief summary, write on the board a segment of DNA bases:

   TAC|AAA|ACA|CGA|CGG|AAT|ACT

   Write the complementary base pairing between DNA and RNA (T—A, A—U, G—C, C—G) on the board. Ask the class to *transcribe* the mRNA complementary sequence:

   AUG|UUU|UGU|GCU|GCC|UUA|UGA

   Divide the mRNA sequence into its triplet codons and convert to the tRNA anticodons:

   UAC AAA ACA CGA CGG AAU ACU

   Rewrite the original DNA base sequence below the anticodon to illustrate that the anticodon is really the same as the DNA message (with the substitution of U for T). Thus, DNA really does directly control synthesis of structural proteins and enzymes. Using the table in Figure 3–24, back *translate* this polypeptide to the anticodon message:

(MET) PHE-CYS-ALA-ALA-LEU (TERM)

h. The result of mitosis must be two identical cells. Identical means genetically identical; that is, each new cell has exactly the same chromosomal makeup as the original cell; there is no variation. Ask the students to come up with ideas about the purpose of mitosis. This list can be hinted at as needed.

   1. growth—we obviously have grown (if not proportionally) from infancy.
   2. development—as we mature, body shape changes as new structures and functions appear in the skin, the sexual organs, and other organs.
   3. replacement—as old cells die, we need a way of replacing them with identical cells.
   4. repair—if new cells couldn't take the place of damaged cells, we'd look like Swiss cheese by the time we reached adolescence!

2. Analogies

a. It serves well to compare the cell to a chemical factory and to consistently draw analogies between the two: from the boss giving orders in the main office, which is isolated from the rest of the factory (the DNA in the nucleus surrounded by the nuclear envelope); to the workers (ribosomes) on the assembly lines in the factory (RER or cytosol) who receive the blueprints from the boss's special messenger (mRNA) and rely on specified raw materials (amino acids) provided by "runners" (tRNA); to the mail room where items are packaged for shipment within the plant or to other factories (the Golgi); to the trams that shuttle the packaged material to the gates of the factory or bring material in from the gates (the vacuoles); to the gates that control shipment in and out (the cellular membrane); to the generators (the mitochondria) that provide the energy (ATP) for all of the factory activities.

b. To convey the look and feel of typical cells, compare them to balloons filled with a colloidal solution of gelatin embedded with fruit cocktail. Although cells are not turgid, they are however deformable, and the organelles are suspended within the cytoplasm just as the fruit is suspended within the gelatin.

c. Compare the phospholipid bilayer to a butter sandwich: The hydrophilic heads are the bread, and the hydrophobic tails are the filling. To achieve the image of a fluid mosaic membrane of phospholipids with proteins and carbohydrates embedded within, create an imaginary visualization of hundreds of ping-pong balls floating on the surface of a pool of water with the occasional larger foam rubber ball interspersed within the continuous sheet of ping-pong balls, and toothpicks are inserted as transmembrane proteins. Frilly deli-style toothpicks can be likened to glycoproteins.

d. Stress that diffusion, osmosis, filtration, and facilitated diffusion are all passive (dissipative) processes. They are driven by dissipating the chemical potential energy stored in a concentration or osmotic or pressure gradient. In contrast, active transport, endocytosis, and exocytosis are all metabolically coupled processes consuming ATP at the same time as the transport event.

   The following analogy may be helpful. Imagine that it's moving day, and the house that you're moving to is just down the hill from the apartment you've been renting. You've boxed up all your belongings, and the boxes are sitting at the top of the hill ready to be carried down. You could just slide them down the hill, relying on potential energy stored originally to get them uphill. This is analogous to simple diffusion. Next, suppose you had two dollies to carry the boxes. This time, the dollies would speed up the transport (facilitated diffusion). Since you have only two dollies, and they each hold only one box, the rate at which you can send boxes down the hill is limited (saturation). Now suppose you discover that one of the boxes you sent down the hill was not your stuff and so it had to be brought back up the hill. You're going to have to get a horse and cart to haul the box back up the hill against gravity (ATP-powered active transport).

e. When introducing the concept of the genetic or triplet code, compare the "reading" of the message of the base sequence in the DNA of each gene to actual reading: the bases equal the letters, the triplet code equals words that mean specific amino acids, and a protein equals the meaning of that sentence after translation. The concept or idea (the protein) can be as short as one sentence or as long as several paragraphs (depending on the size of the gene). Reading the message is like reading letters on a page made of words all three characters long and with no space between words. If we change the order of the words, we get a different meaning. For example, "The man sat on the hat," has a very different meaning than "The hat sat on the man," even though both sentences use the same letters and words. In addition, we respond to "signals" when we read: A capital letter means the beginning of a sentence; a period tells us that the sentence is over. Punctuation marks act just as the regulatory triplets within each segment of a gene.

f. The letter-word-sentence analogy can be continued to illustrate point mutations and frame-shift mutations. Consider the message (DNA) "I looked all over for the cat." If a mutation occurs at a specific letter (nucleotide) changing the "t" in cat to an "r," the meaning changes completely. The crippling genetic disease *sickle cell anemia* results from the *mutant* substitution of a *single* nucleotide! A *deletion* mutation of a base that results in a *frame-shift mutation* can be just as problematic. Consider the triplet code message: "Theredhatforthemanwastoobig." Separated into triplet codes it reads, "The/red/hat/for/the/man/was/too/big." But if one nucleotide were deleted (for example, the "r" in red), the message would read: "The/edh/atf/ort/hem/anw/ast/oob/ig." Such a nonsense sequence would likely create a useless protein.

3. Demonstrations

a. Identifying the different mitotic stages through projection slides in lecture or lab will help students develop their abilities to make sense of what they see through the microscope. Present images of various mitotic stages and have the students arrange them into their logical temporal sequence.

b. The concept of "increased surface area" and its structural/functional importance is often difficult for students to fully appreciate. To demonstrate this concept, lay a piece of notebook paper on a table, and point out that the paper is occupying an $8.5'' \times 11''$ (93.5 sq. inches) space on the surface of the table. Now if you accordion-pleat the paper, you can illustrate that the same $8.5'' \times 11''$ sheet will occupy a space of about $8.5'' \times 3\ 5.5''$ (46.8 sq. inches). In organs, such as the absorptive lining of the small intestine, the greater the surface area, the greater will be the amount of transport work that can be done. By increasing the inside surface area of the intestine with plicae, villi, and microvilli, the absorptive surface area has been increased many times more than a smooth-walled tube of the same outside diameter would provide.

c. A commonly used, if indelicate, mnemonic device to help remember the major stages of the cell's cycle is "I Pee on the MAT." (Interphase, Prophase, Metaphase, Anaphase, Telophase). PMAT—"Pour Me Another Tea" is an alternative.

d. A stem cell gets its name, because it's the beginning of a line of cellular growth, like the main stem of a plant that eventually grows (differentiates) into branches, leaves, and flowers. Ask them to imagine how many branches the simple spherical fertilized egg moves along to form the human body consisting of thousands of different cell types.

e. The weakness of hydrogen bonds makes them something like Post-It Notes, and this is an important aspect of their structural/functional role. They stick well enough but are easy to lift when necessary. Because they are weak (about 1 kcal/mole) compared to thermal energy *k*T, (about 2.3 kcal/mole) they are very easy to break. In the function of DNA, for example, it needs to be unzippered for transcription or replication to occur. The H-bonds between complementary bases are only two or three in number and so average about 2.5 kcal/mole. This accounts for the ease with which unzippering occurs: it's scarcely different from *k*T.

4. Common Student Misconceptions/Problems

a. Students may find it difficult to see how the organelles fit together to form a cell. A three-dimensional cell model, available from most lab supply firms, that illustrates the "typical" animal cell can help overcome this problem. Have students describe the processes of transcription and translation and secretion as they make physical reference to the model. Ask them to find the ATP-generating organelles.

b. Another gedanken experiment. Trying to get the class to think concretely about an important physical process.

How Do These Factors Influence Diffusion Rate?

Factors	**Effect** (faster/slower)
Distance	
Size of the gradient	
Molecular size	
Temperature	
Viscosity	
Electrical forces (same vs. opposite)	

In lecture or lab, make a table on the board, and ask the students to consider each factor in turn. Ask them to predict how each would affect (+ or −) the diffusion rate and to say why they think that. This exercise should help them develop a physical intuition about unseen molecular processes. Accumulate the *correct* responses on the board. Guide them toward those.

More gedanken: Equilibrium does not mean that diffusion does not continue, only that there is no longer an electrochemical gradient to drive transport and hence no **net** movement. Molecular motion, which underlies diffusional transport, never ceases. Have them imagine how a cluster of solute molecules would be dispersed in a "random walk" as they are buffeted about by collisions with moving solvent molecules. Look at computer animations of diffusion. In the lab, have students use a microscope to observe the random movement (*Brownian motion*) of oil droplets in very dilute milk or suspended Lycopodium powder mounted under a cover slip.

c. Make clear that the words *isotonic*, *hypotonic*, and *hypertonic* refer to the behavior of a cell exposed to a solution and not specifically to the physical-chemical properties of the solution. By definition, a solution is *isotonic* for a particular cell if that cell *neither swells nor shrinks* when placed in the solution. An *isosmotic* solution will not be isotonic if it is made with a solute that permeates the plasma membrane, that is, cannot exert a steady osmotic pressure to balance the osmotic (swelling) pressure of the cytoplasm. [Aside: One of us (adm) measured the swelling pressure of the myofilament lattice of skeletal muscle by x-ray diffraction and secondary osmometry. It is about 30 torr (1 torr = 1 mmHg), that is, about the same as the colloid osmotic pressure of plasma.]

d. It is common for students to confuse flagellum, cilium, and microvillus. Have the students do a summary table (in class, in lab, or at home) comparing each structure's appearance, relative size, density, structural makeup, and function with an example of each, to alleviate some of the confusion. Flagella are long, whiplike appendages composed of microtubules (in eukaryotes). The cells use them to propel themselves from place to place (locomotion). The sperm cell is the only human cell with a flagellum. Cilia are small, hairlike projections on the surface of the cell, also composed of microtubules. Because they are so numerous, they often give the cell a "furry" appearance. In the living state, they beat, or wave, in a coordinated manner to sweep extracellular substances across the cell's surface. Example: cells that line the airways or the uterine tubes. Microvilli are finger-like projections from the cell's membrane that enormously increase surface area. They are supported by microfilaments, not microtubules. They do not exhibit the characteristic whiplike movements of cilia and flagellae.

e. The term *interphase* is somewhat misleading since it implies a short phase stuck in between other activities. Point out that this couldn't be further from the truth; interphase comprises most of the cell's life span, even in cell populations with very high mitotic rates.

5. Vocabulary Aids

a. Be careful to distinguish between *semipermeable* and *selectively permeable.* Any physical barrier, living or nonliving, can be semipermeable (i.e., permeable to some molecules but not others) by virtue of the size of its pores. Living membranes can be selectively permeable, because they can change their permeability and select for or against a substance by opening or closing or inserting a channel or activating a transporter.

b. When defining the prefixes *iso-*, *hypo-*, and *hyper-*, associate the "s" in iso with the "s" in the word "*s*ame." For *hypo,* use the example of *hypo*dermic, below the skin. For hyper, suggest *hyper*active.

c. Many of the terms for cellular organelles end with the *suffix* "–some." Ask the students to make a list (help them get to chromosome, nucleosome, ribosome, lysosome, peroxisome, proteasome, centrosome) and then to look at the *prefix* to find the meanings. Besides demonstrating suffix and prefix as essential elements in medical terminology, this will help later when they encounter "soma," the neuron cell body.

d. As an aid to memory for the pairings of nucleotides, think of "AT" as being where it's @. The notion is to couple two essential components of modernity, DNA and all that it promises and threatens (AT) with the hipness of the Internet email address symbol. GC is all that's left to remember. [: >.}

e. *Guanine* (a purine) is named for guano, bird droppings, from which it was purified in 1850. *Adenine* (the other purine) is named for *adeno-*, the prefix for gland. It was first crystallized from pancreas in 1885.

f. To remember that thymine (T) in DNA is replaced by uracil (U) in RNA, recall that if you order it, "Tea goes to you." T —> U

CHAPTER 4

The Tissue Level of Organization

Introduction

After defining *tissues* as an anatomical concept, you might start the discussion of tissues by asserting that cancers of muscle and nerve are relatively rare whereas cancers of bone marrow, gut, and skin are far more common. Ask your class: Why do they think this happens? What is it about their tissue components that make some organs more prone to cancer than others? Guide them by asking, When do mutations get passed to the next cell generation? Answer: during cell division. What condition requires rapid cell division? We are composed of about 200 different cell types. By combining them in many ways, organs can be endowed with a wide range of abilities (and limitations too, of course). Students are usually familiar with **cells** and **organs** but usually not with the **tissue** level of organization. Point out that different organs sometimes need to accomplish similar functions and so will contain similar tissues. For example, the heart, the gut, and the uterus all need to generate force, and so all contain muscle tissue, although of three different types. Gut also needs to secrete and absorb, and so is lined by a versatile simple columnar epithelium with mucous cells. Also, all organs need tissues specialized to hold the tissues together and impose a basic shape.

Instructional Goals/Learning Outcomes

1. Identify the four major types of tissues in the body and describe their roles.
2. Discuss the types and functions of epithelial tissue.
3. Describe the relationship between form and function for each type of epithelium.
4. Compare the structures and functions of the various types of connective tissues.
5. Describe how cartilage and bone function as a supporting connective tissue.
6. Explain how epithelial and connective tissues combine to form four types of membranes, and specify the functions of each.
7. Describe how connective tissue establishes the framework of the body.
8. Describe the three types of muscle tissue and the special structural features of each type.
9. Discuss the basic structure and role of neural tissue.

10. Describe how injuries affect the tissues of the body.
11. Describe how aging affects the tissues of the body.

Teaching Strategies

1. Lecture and Lab Ideas

a. Exploit light microscopy of the amphibian (salamander) gut tube in cross section (widely available from biological supply houses) as the model organ for introducing tissues and, from that, organ building. The four tissue types and two membrane types are readily demonstrated, and how they interact to accomplish organ function (absorption, secretion, motility, enzymatic digestion) can be naturally presented and discussed. Point out the mucosa that lines the *lumen* (a word they need to hear now and often). Use high power to show the simple columnar epithelium with mucous cells intercalated. Talk about exocrine and endocrine secretion (enzymes and mucus; gut hormones) by different cell types. Mention the passive and active uptake of nutrients by this epithelium too. Relate these to the molecular physiology of Chapters 2 and 3. Point out the presence of immune cells in the connective tissue to guard against invading pathogens. Discuss the sensory and motor nervous tissues in the gut wall. Describe the smooth muscle layers (ask them why two muscle layers are at right angles?), and finally the serous (watery) membrane of the mesentery that *lubricates* the exterior of the organ and protects against *adhesions* among the viscera.

b. As you present each type of specialized cell junction, emphasize the "logic" of the structure related to its function. In lab, have the students draw each junction type relying on Figure 4–2, describe its function, and give an example of location.

   1. Cell adhesion molecules (CAMs) are transmembrane glycoproteins that attach large areas of opposing plasma membranes. For example, CAMs attach the base of epithelia to the underlying basement membrane.

   2. Gap junctions consist of pairs of integral membrane proteins that "snap together" between adjacent cells to form channels ("connexons") that allow small molecules (up to about 1200 daltons) to pass directly from cell to cell. Gap junctions are common in certain epithelia and in cardiac and smooth muscle tissues, where they allow the spread of electrical excitation from cell to cell.

   3. Tight junctions are formed by fusion of neighboring plasma membranes by integral membrane proteins. They limit paracellular transport of small molecules, including water, across an epithelium. Tight junctions exist between cells that line the gut lumen.

   4. Desmosomes are like spot welds that anchor two cells together at a point. They are formed by CAMs extending from both cells; intermediate filaments attach to the cytoplasmic side. They are especially abundant in skin epithelium, giving it mechanical strength.

   5. After describing the cell junctions one by one, remind the students that commonly more than one type of junction may link neighboring cells. A good example of this are the intercalated discs that join cardiac muscle cells. They contain both gap junctions (to allow synchrony of electrical activity) and desmosomes (to transfer force from cell to cell).

c. Classification of epithelia: With the description of each type of epithelial tissue, emphasize the "logic" of the structure related to its function. In lab, ideally from their own microscopic examination, have the students draw each type, describe its function, and give an example of location. To encourage concrete and specific learning, require that the example be specific; for example, do not accept "lung" as a location for simple squamous epithelium, since many kinds of tissues form the lung. "Lining the alveoli within the lung" would be more satisfactory.

d. Strive to help the students appreciate the ubiquity and variety of connective tissues while showing how each adheres to the basic plan (cells + matrix; matrix = ground substance + fibers). With the description of each type of connective tissue, emphasize the "logic" of the structure related to its function. In lab, have the students draw each type from examination of micro slides if possible, state the basic plan components, describe the tissue's functions, and indicate what organ the slide came from. As with epithelia, make sure that the example is specific. The matrix of cartilage is diverse; it may be firm enough to support weight (think of articular cartilages), or it may spring back after being bent (think of the elastic cartilage of the pinnae). Cartilage is classified according to the type and proportion of fibers in the ground substance. Because of the scant collagen in the transparent matrix, hyaline cartilage appears glassy (hence the name). The cartilage at the end of a chicken breast bone is an example of hyaline cartilage with which most students are familiar. Elastic cartilage contains abundant elastin fibers, whereas the tough, stiff fibrous cartilage is loaded with collagen.

e. Muscle tissue is the only tissue capable of contracting suddenly when appropriately stimulated. The cells of muscle tissue are very different in appearance from epithelial or connective tissue cells; they are long and fibrous. In lab have the students draw each type, describe its function, and give an example of location. As with epithelia and connective tissues, make sure that the example is specific.

f. As with the introduction into the other tissue types, remind the students that neurons are distinctive, because they can convert an electrical impulse into a chemical signal that is used to control other cells. Point out that this occurs at the synapse. In lab, have the students draw a typical neuron and label and identify the function of principal parts.

g. Don't forget to refer students to the extensive preview of organ systems in Figure 4–2 (a) to (l). It offers a concise summary of each organ system utilizing both simple summary drawings and a condensed narrative in tabular form.

2. Analogies

a. A parallel can be drawn between the cells and tissues within an organ, and the instruments and sections within an orchestra. An orchestra is made up of many individual instruments, but they are not randomly put together. They are grouped into sections according to the sound and construction of the instruments: the strings, the winds, the brass, and the percussion. Each instrument (cell) is playing in harmony and synchrony (tissue), and with the help of a conductor, the individual sections work together to make one large, harmonious orchestra (organ).

b. To get truly "inside" the body, it is necessary to cross an epithelium. Thus, food molecules within the lumen of the gut are not truly *inside* the body. The epithelial cells must transport the food substance across the epithelium before it is literally inside.

c. Reticular connective tissue fibers form an open mesh like a furnace filter. Immune cells cover the fibers, and as lymph percolates past them, the immune cells can grab and trap pathogens and foreign antigens.

d. Glandular epithelia: Stress that the key difference between endocrine and exocrine glands is like the difference between companies that produce for export only and those that produce for domestic consumption. In other words, what matters is which *direction* the newly manufactured "goods" move. An endocrine cell releases its product across the basal lamina *into* the body; an exocrine gland sends its product *outside* the body by releasing it across the apical plasma membrane into a duct or onto surface. A duct is like a pipeline. Rely on Figure 4–6 in lecture or lab to present the three forms of exocrine secretion. Be sure to point out that endocrine secretion is almost always (except for lipophilic hormones) merocrine, that is, by release of vesicles containing the hormone. Point out its similarity to synaptic release of neurotransmitters from axon terminals.

e. An epithelium is like an open-face sandwich. It's polarized, with a definite top and bottom.

f. The movement of a particle along a sheet of mucus on a ciliated epithelium is like handing off a hapless student over the heads of the crowd at a rowdy football game.

3. Applications

a. Advertisements for many skin creams claim to "replenish" collagen and elastin. Do your students believe that putting such molecules on the epithelium can cause them to be incorporated in the collagen and elastic fiber in the underlying connective tissue? Where do these fibers actually come from? Fibroblasts and fibrocytes. Not from the epithelium!

b. Keratin, which is characteristic of epithelial cells but not of immune cells, can be looked for in biopsied lymph nodes. If keratin is detected, it may be a sign that a cancer of epithelial origin (most are) has spread outside the original tumor.

c. Drugs that affect cell adhesion molecules are being tested to treat disease of the immune system and slow the spread of cancer.

d. The polarity of epithelial cells results partly from the ability of the Golgi apparatus to target proteins for delivery to particular "addresses" in the cell.

4. Common Student Misconceptions/Problems

a. From previous biology classes and Chapter 3 in this text, most students have an adequate concept of a cell. Most could identify or give examples of several organs. It is the *tissues*, the level of organization between cells and organs, where their knowledge is limited. (For many, it will be the first time the word *tissue* doesn't mean something into which you blow your nose!) By employing the primitive gut tube as the archetypical organ built up from multiple concentric layers of different tissues, the concept of tissues cooperating to produce a functioning organ can be made clear. Secretion, absorption, blood flow, sensory and motor nervous function, immune defense, and movement are functions naturally motivated by a systematic review of the tissue components in this organ. Emphasize that tissues are the "building blocks" of organs; only by understanding the tissues can one develop a clear understanding of organs.

b. Despite the great diversity of organs, if they were disassembled into tissues, and the tissues were sorted into "piles" based on their basic function, there would be only four groups! This abstraction that everything in our body can be categorized into just four tissue groupings can be baffling to students given how apparently different a brain is from a heart, or a stomach is from a muscle. The resolution to this apparent paradox is simply that the various tissues that we sort into the four piles are actually quite diverse (e.g., consider bone and blood!). To help them grasp this concept, analyze why bone and blood are both connective tissues in terms of cells (osteocyte; RBC), ground substance (hydroxyapatite; plasma), and fibers (collagen; fibrin).

c. Although connective tissue may seem to be a conglomeration of unrelated tissues, point out that they all share a common plan: a noncellular matrix that supports the cells that made the matrix and that largely determines the mechanical properties of the tissue. The concept of a *noncellular* matrix may be difficult for the students to grasp, since they have always been told that all living things are made of cells. True, but only part of the story!

5. Vocabulary Aids

a. Stem cells are cells found at the base of an epithelium and retain the ability to divide; that is, this is where the more superficial cells *stem* from.

b. Squamous epithelia: Associate the word "**squa**mous" with "**sca**ly," which is the Latin root from which the word squamous derives.

c. Associate the word "*strati*fied" with layers; associate it with the *strato*sphere, the middle *layer* of the earth's atmosphere, or with *strata*, the *layers* we see in the earth's crust.

d. "Ciliated pseudostratified columnar epithelium" is a tongue-twister that will trigger a "skip over" response in many students. (Sadly, this is true for many terms presented in this text.) The best way to lead them to understanding this term and other combinations is to put the whole darn phrase on the board and take them through each term and then rehearse them in saying it as a group. Tell them this is how to deal with complicated anatomical or physiological terms. DON'T SKIP THEM. Teach them how to dissect them, say them, and understand them. And say them again and say what they mean. This may be the most important dissection skill they acquire in the course, to peel apart complex unfamiliar terms into their constitutive prefixes, roots, and suffixes. To function professionally, one must both walk the walk and *talk the talk.*

e. Associate the word *lacuna* with *lake* (to which it relates). Lacunae look like little lakes (clear areas) in sections of cartilage and bone under the microscope. In the lacunae are found the chondrocytes or osteocytes, respectively.

f. Canaliculi means "little canal"; it brings to mind a type of tubular pasta (cannelloni).

g. Some students may confuse "stratified" and "striated." Saying them out loud will help since they only look alike but sound a lot different. Remind them of the meaning of stratified and that it refers to the organization of certain epithelia. Associate the sound of the word ***stri**ated* with the sound of the word ***strip**ed.*

h. Stress that the direction of information transfer is one way: from dendrites through the soma and to the axon. Associate the "**t**" in dendrite with "**t**o" and the "**a**" in axon with "**a**way."

i. *Mero-* means part, as in polymer. *Mero*crine secretion releases only part of what is there. *Apo-* refers to apex. *Apo*crine secretion involves release of the apical portion. *Holo-* refers to whole. In *holo*crine secretion, the entire cell is shed.

j. Mucus is a noun; mucous is an adjective. *Mucous* cells release *mucus*, a solution of *mucins* in water.

CHAPTER 5

The Integumentary System

Introduction

Students may not consider the skin to be an interesting organ compared to the heart or the brain. However, on a day by day basis, most of us are far more concerned with our skin than with any other organ. Common concerns include psoriasis, rosacea, skin infections such as acne, allergic reactions, cuts, bruises, abrasions, insect bites, and even aesthetic worries about wrinkles. Our culture's near obsession with a "healthy tan" has led to a pandemic of skin cancer. The incidence of skin cancer has increased dramatically in the past 50 years. Of the several forms of skin cancer, one is particularly dangerous, a *melanoma*. A melanoma that has metastasized (spread from its initial site) is often fatal. It is now widely accepted that the primary cause of these cancers is the invisible, short-wavelength **ultraviolet** component in sunlight. How UV light induces mutations and cancer in sun-exposed skin is interesting, as are the means by which the skin defends itself against UV injury (e.g., melanin synthesis, intercellular melanin transfer, enzymatic DNA repair). For a popular account of many fascinating aspects of the biology of the skin, see Marc Lappe's *The Body's Edge: Our Cultural Obsession with Skin* (1996, Henry Holt and Company).

Instructional Goals/Learning Outcomes

1. Describe the main structural features of the epidermis, and explain the functional significance of each.
2. Explain what accounts for individual differences in skin color, and discuss the response of melanocytes to sunlight exposure.
3. Describe the interaction between sunlight and vitamin D3 production.
4. Describe the roles of epidermal growth factor.
5. Describe the structure and functions of the dermis.
6. Describe the structure and functions of the hypodermis.
7. Describe the mechanisms that produce hair, and explain the structural basis for hair texture and color.
8. Discuss the various kinds of glands in the skin, and list the secretions of those glands.
9. Describe the anatomical structure of nails, and explain how they are formed.
10. Explain how the skin responds to injury and repairs itself.
11. Summarize the effects of the aging process on the skin.
12. Give examples of interactions between the integumentary system and each of the other organ systems.

Teaching Strategies

1. Lecture and Lab Ideas

a. Take this opportunity to specifically characterize each tissue component in skin as being an example of tissue categories encountered in the previous chapter. A cornified stratified squamous epithelium, a dense irregular connective tissue dermis, delicate slips of smooth muscle between hair follicle and epidermal–dermal boundary, exquisitely sensitive hair mechanoreceptors—useful examples abound.

b. Linger on Figure 5–1. It's jammed with rich anatomical detail and offers a good chance to help the class develop their skills at viewing, understanding, and learning anatomy from drawings and other figures. Tell the class explicitly that you are trying to help them with this difficult but needed cluster of skills. Try a simple ask-and-respond strategy coupled with projection of the image or looking at the text figure, if projection is not possible. Ask, How many layers to the skin? All kinds of numbers may be forthcoming but insist that they carefully study the diagram to see how many layers are identified. Two. Epidermis, dermis. What about hypodermis? If not skin, then what is it? Look at the epidermal–dermal boundary. What does all that folding get you? More adhesive surface for molecular glue. Point out the capillary loops for heat dissipation, and the A-V shunts in the hypodermis.

c. As phagocytic and antigen-presenting cells, dendritic cells can initiate an immune response by intercepting and processing invading bacteria. Underscore the functional wisdom of locating these cells in the stratum spinosum, superficial to the basal lamina, to provide an early warning to the defense system.

d. As keratinocytes leave the stratum germinativum, and enter the stratum granulosum and then push up layer by layer into the stratum corneum, they lose their nuclei and other organelles, fill with keratin granules that form keratin fibers, and then die to be shed as scaly plates. These scaly keratinocytes form a "suit of armor." The resulting cornified water-resistant barrier was an important evolutionary development that allowed amphibians to adapt to terrestrial habitats.

e. The stratum corneum varies in thickness depending upon the location of the skin and pressure or friction on the skin.

f. Emphasize that it is as important to keep body water *in*, as it is to keep environmental water *out*. Explain the several mechanisms skin uses to limit water movement across it. Then, point out that sweating represents a deliberate *loss* of body water, the demands of thermoregulation having a higher priority than water conservation. If not replaced, cutaneous water loss can be fatal.

g. Although the dermis is composed of two layers, the demarcation where the papillary layer ends and the reticular layer begins is not very distinct. There is a gradual change from looser connective tissue in the papillary layer (where delicate blood vessels and sensory nerves are located) to the dense irregular connective tissue of the reticular layer beneath it. Here lies most of the skin's toughness and elasticity.

h. An individual hair follicle is an invagination of the epidermal surface; hence, the germinative cell layer of the epidermis can be followed down into the dermis where it forms a germinative layer for the hair follicle.

i. Draw attention to the structure of the sebaceous and sweat glands. As exocrine glands, they possess *ducts* that lead to either a hair follicle or toward the skin's surface.

j. As far as skin color goes, all humans (except for albinos) are different shades of brown. We are all tinted with melanin, some of it reddish brown, some of it purplish brown, some of it brown black, all of it brown. Just like dog lips.

k. Emphasize that the skin plays a crucial role in balancing heat loss with heat gain. A constant internal temperature is a crucial aspect of homeostasis.

2. Analogies

a. As a way of illustrating that the cells of the epidermis originate in the stratum germinativum and are pushed to the surface from below, remind the class about how grass grows. The pale tender part next to the roots is where mitosis occurs. The cells are then pushed up toward the top, elongate, acquire organelles, and mature. At the tip, they die and turn brown. The lawn mower trims the dead cells off. This is like desquamation.

b. If you could peel the epidermal layer off the dermis, the papillary layer of the dermis would resemble an "egg crate" mattress. [Refer to Figure 5–2(a).]

c. If a wound doesn't penetrate the basement membrane, the skin will heal without a scar. This is like replacing peeling paint: The underlying wood grain (dermal structure) is preserved. But if the wound breaks into the dermis (blood loss signals this), healing can occur but the structure of the dermis will be replaced with random components. This is like filling a hole in wood with putty.

3. Demonstrations

a. The secretion of watery sweat by merocrine glands in the skin is turned on by the sympathetic nervous system. Pain usually causes generalized sympathetic activation. Have students look at their fingertip under a dissecting microscope, and then observe the response to mild pain by gently biting a finger of the opposite hand (activating cutaneous nociceptors). They will see beads of transparent watery fluid appear within a few seconds of stimulation. They will be amazed.

b. To emphasize the danger of UV light to the skin, point out that commercial tanning beds strongly resemble caskets! The American Medical Association has called for a ban on tanning equipment.

c. To see the contribution of hemoglobin to skin color, look at the skin underneath a fingernail. (Can't wear polish for this.) Now press firmly on tip of nail to push blood away from surface. Wait until steady. Now let up and watch the capillaries refill with blood. They can use this trick to look for jaundice in the skin of a dark-skinned person (also look in the mouth and at the sclera).

4. Applications

a. Psoriasis results from an increase in the mitotic rate of the stratum germinativum causing the epidermal layer to be replaced within a week instead of the normal 2–4 weeks. Many of the treatment strategies target mitosis to produce clinical improvement.

b. After a long bath, our fingers resemble prunes because of the swelling that occurs as the hypotonic bath water flows osmotically into the cells of the epidermis. Because of anchoring to the dermal boundary, the epidermis is forced to fold to accommodate the extra cell volume.

c. Ironically, the "tan" that we find so attractive (caused by increased melanin synthesis by melanocytes responding to the UV light) can also fragment DNA. This damage can result in epidermal tumors derived from basal cells (basal cell carcinoma), or from more superficial cells (squamous cell carcinoma), or from melanocytes (malignant melanoma). Since the basal cells and squamous cells tend to remain within the epidermis, they are less likely to metastasize. If removed early enough, they can be cured. However, melanocytes are intrinsically invasive, and soon leave the epidermis to enter the dermis. From there they can easily enter blood and lymph vessels and spread to distant organs such as the brain, liver, and lung. Consequently, basal and squamous cell carcinomas are rarely fatal, but malignant melanomas unfortunately often are.

d. Dermabrasion is a process that uses a spinning metal brush to remove the epidermis and upper layers of the dermis in order to smooth disfigured or scarred skin. The dermatologist must be careful not to sand down deeper than the epidermal derivatives. If tissue removal goes too deep, epithelization won't occur and extensive scarring will result.

e. A *birthmark* can refer to a hyperpigmented skin feature (*nevus* or mole) or result from a vascular problem. "Port wine stains" (formerly called capillary hemangiomas) result from dilated blood vessels that lack a nervous supply. They can be faded with laser treatment but often reappear.

f. In spite of what commercials for hair products may lead you to believe, the hair shaft itself is dead; it cannot be nourished once it has formed. The composition and structure of hair reflects physiological factors that affect the active cells of the follicle. Sometimes, bodies will be exhumed, even years after death, for the purpose of a forensic examination of the hair to see if there is residual evidence of poisoning.

g. If an injury involves a first- or second-degree burn, elements of the epidermis and dermis remain to replace the destroyed cells. If an injury produces a third-degree (or full-thickness) burn, there is no skin tissue left and so no capacity for mitosis and cell replacement. In this case, skin grafts must be used to repair the damage.

h. The use of tissue culture to grow giant sheets of artificial epidermis from the patient's own skin or from donor skin (neonatal foreskins are often used because the cells grow well in tissue culture) represents the application of advanced cell biology to the care of badly burned children and adults.

5. Common Student Misconceptions/Problems

a. Students may consider the integumentary system to be merely the visible epidermal layer. Point out that the epidermis is really quite thin (only 50 μ in some places). Explain that the dermis is rather thick (it is what leather is made from) and that the skin includes all four tissue types, and thus is an organ. Because of the skin's essential interactions with the rest of the body along with the presence of organs such as hair follicles, it truly is an organ *system*.

b. For some reason, students often confuse keratinocytes and *keratin* with melanocytes and *melanin*. Emphasize that the former are the predominant cell type of the epidermis and produce keratin that contributes to the structure of the epidermis, hair, nails, and tooth enamel, that is, where a need exists for a hard, tough, water-resistant material.

Melanocytes, in contrast, are specialized cells in the stratum germinativum that produce melanin, a dark pigment that protects from UV light.

c. Even though the basal cells are *cuboidal*, the entire epidermis is classified as stratified *squamous* epithelium, because that is the cell shape of the most superficial (apical) cell.

d. Students commonly assume that the dermis must give rise to the hair follicles or sweat glands, since that is where most of the structure is found. To counter this misconception, reiterate that they are called "epidermal derivatives." In lab, use a skin model or large diagram or micrograph to show the direct cellular continuity between the stratum germinativum and these structures. Also, make the point that these epithelial structures could not have been formed from connective tissue cells.

e. Help the class understand that not all stratified squamous epithlia are cornified. They may confuse stratum *corneum* with *cornified*. When moist, the superficial layer of a mucosa such as anus, vagina, oral cavity, or esophagus will not have horny or keratinized cells. If you say "horny" in class, many will laugh. Your choice.

6. Vocabulary Aids

a. Compare the word ***stratum*** to the already familiar word ***strat****ified*, and indicate that the root *stratum* means "a layer."

b. Associate the word *germinativum* with "germinate." This will emphasize that this population of cells can divide, and thus grow, just as seeds that *germinate*.

c. The cells of the *stratum spinosum* get their name because of their "spiny" appearance. The spiny cell margins result from tissue shrinkage that reveals the abundant desmosomes that link the cells together.

d. The keratohyaline granules within the cells of the *stratum granulosum* give them a "granular" or "grainy" appearance.

e. The keratin of the stratum *corn*eum makes it as tough as a rhino's horn. The words *horn* and *corn* are related etymologically. (Feel free to make bad jokes about the word *horny*.)

f. Nicotine or estrogen patches are examples of *transdermal* medication delivery systems; they are most useful with lipid-soluble drugs such as these.

g. Associate the word *vellus* ("peach fuzz") with "velvet" (to which it is related).

h. *Papilla*, as in papillary layer of the dermis, refers to the Latin word for nipple. Many things remind anatomists of nipples.

CHAPTER 6

Osseous Tissue and Bone Structure

Introduction

Concern about osteoporosis provides a useful way to motivate interest in the subject of bone biology. Begin with data on osteoporosis prevalence; note that it affects women far more often than men. Is the prevalence increasing? Next, solicit input from students concerning relatives and others they know who have been afflicted. Ask about signs and symptoms (fracture, pain, deformity). Do more females suffer than males? As examples of interactions between the skeletal system and other organ systems, point out the impact on bone of estrogen, a female reproductive hormone; vitamin D from the integumentary system; calcium and phosphate from the digestive system; and load-bearing exercise neuromuscular system. Promise the class that this unit will help them understand how to promote bone health, and how the modern therapies to prevent and treat this common crippling disease work in terms of cellular targets. You can learn more about osteoporosis from a review in the March 2003 issue of *Scientific American* by Clifford J. Rosen entitled, "Restoring Aging Bones."

Instructional Goals/Learning Outcomes

1. Describe the primary functions of the skeletal system.
2. Classify bones according to shape and internal organization, giving examples of each type, and explain the functional significance of each of the major types of bone markings.
3. Identify the cell types in bone, and list their major functions.
4. Compare the structures and functions of compact and spongy bone.
5. Compare the mechanisms of endochondral ossification and intramembranous ossification.
6. Describe the remodeling and homeostatic mechanisms of the skeletal system.
7. Discuss the effects of exercise, hormones, and nutrition on bone development and on the skeletal system.
8. Explain the role of calcium as it relates to the skeletal system.
9. Describe the types of fractures, and explain how fractures heal.
10. Summarize the effects of the aging process on the skeletal system.

Teaching Strategies

1. Lecture and Lab Ideas

a. When describing the role of PTH and calcitonin in the control of blood calcium, begin by diagramming the general scheme for a negative feedback control system:

(disturbance) > RECEPTOR > CONTROL CENTER > EFFECTOR > (response)

Leave this scheme up, and then show Figure 6–16, using the scheme to conceptualize the information in the text figure.

Don't forget to add and label the feedback path that "feeds" the corrective action "back" to the disturbance, that is, a departure from the normal range of the physiological variable that is being controlled. In this example, that variable is the concentration of ionized calcium in the plasma. First, for factors that increase calcium (panel a) ask them to systematically match up the real-world components in the projected figure with the theoretical scheme above, beginning with the disturbance. They may have a hard time acknowledging that the receptor and control center are both located in the same place (parathyroid gland cell). Repeat the exercise for panel b. To keep the discussion realistic, point out that it doesn't come into play until calcium ion levels have risen to levels not often reached in healthy persons.

b. In your opening remarks, tell the class there are five basic functions of the skeletal system. Ask them to think what they might be. Write them on the board as they suggest them. After "support" they will likely be stymied to name others. After you complete the list, this exercise will teach them that they know some stuff but have much to learn!

c. The skeletal system includes more than bones; it includes joints, ligaments, cartilage, dense regular connective tissue, and joint capsules too.

d. When presenting the bone shape categories, name a type (e.g., long bone) and then ask the class to come up with an example from their general knowledge. Give them hints for sesamoid, etc. as needed. This exercise will ideally condition them to offer in-class suggestions and help persuade them that they know something of the topic already.

e. Make special mention of "mesenchyme" when presenting the process of intramembranous bone growth. Let the class know that this spidery, stellate cell from the mesoderm of the embryo is a key stem cell for a wide range of mesodermal derivatives (bone, cartilage, fibroblasts, muscle). This anticipates the discussion of germ layers in the final chapter, Chapter 29, of the textbook.

f. Project Table 6–2. In part, the information anticipates Chapter 18 but point out to the students that at least some of these hormones are *already* known to them. This table of hormones dramatizes the complexity of organ regulation and the maintenance of homeostasis.

2. Analogies

a. One analogy to present is a *backwards* one, that is, what bone is *not* like. It is not like wood or stone. Is the skeleton similar to the framing of a house? Are the joists and piers like the ribs and femurs? If the load in a house increases, the floors and roof simply sag. If it were a skeleton formed of living, responsive bone, as the load increased (e.g., due to weight gain or bone-loading exercise), the bone would respond by depositing additional load-bearing matrix, rather than sag. You can use this same idea with stone buildings, such as the Colosseum in Rome, just to exploit a dramatic and familiar example. One hardly expects the stone arches (the *trabeculae*) to become stouter under the popular parts of the stadia!

b. A man-made example of a structural analogy to the functional histology of bone is fiberglass-reinforced resin used to build cars, boats, skis, and surfboards. Just as bone does, these objects need to combine compressive strength with flexibility. The collagen fibers act like the glass fibers to give flexibility, while the mineral matrix of calcium phosphate crystals supplies the rigidity to support an axial load. A related analogy is to structures built from steel reinforcing rods (*rebar*) and concrete.

c. Suppose you were an engineer commissioned to design a bone. Functionally, bones need to be strong to protect organs and support weight. A solid, cylindrical shape would work but would waste material and be heavy. And so the engineer would quickly realize that a pipe-like hollow cylinder would not only have almost as much strength as a solid column, but the hollow interior provides a space to accommodate cell activities (blood formation in the red marrow) but also energy storage (adipocytes in the yellow marrow). Next, the engineer would devise ways to repurpose familiar ingredients (osteoblasts, mineral, collagen) into two mechanically and spatially very different tissues, compact and spongy bone. What a dandy engineer you are!

d. The concentric lamellae of an osteon look like the rings seen in a cross section through a tree trunk. It is an apt analogy since osteons grow appositionally layer by layer just as trees do. In some animals, cross sections through the teeth reveal layers corresponding to the age of the animal.

e. The fine cell processes of the osteocytes as they extend radially through the canaliculi to touch and spread nutrients are reminiscent of the extended fingers in the "Creation of Adam" panel in the Sistine Chapel ceiling.

3. Demonstrations

a. A dramatic demonstration can be performed to illustrate the weight-bearing properties of a cylinder. You will need an 8.5 × 11-inch sheet of paper, a paper plate, some tape, a scissors, and a rock. Ask the students if they ever played the childhood game of "rock-paper-scissors" and remind them that "scissors cut paper, paper covers rock, and rock breaks scissors." While holding the piece of paper on its edge, announce that you are going to prove that paper is stronger than either scissors or rock by balancing both on the paper's edge. Roll the paper lengthwise, and tape the short edges together, so that you form a cylinder. Stand the cylinder on the table, and place the paper plate on top. Now place the scissors and rock on the plate. Voila! Now discuss the analogy between the tubular shape and the "engineering" of long-bone shape.

b. To demonstrate the many functionally important and informative surface features of bone, there is nothing better than opening a "bone box" in lab and pulling out one beautiful, dramatic, REAL human bone after another. Getting them to name them is easy, and point out the radial **tuberosity**, the **foramen** magnum, the frontal paranasal **sinus**, the **facets** of the skull and the atlas, lateral and medial **condyles**, and so on. If the cortical bone is broken over the epiphysis as it often is, show the students the delicate trabeculae, bony arches, thus revealed.

c. A lecture or lab demonstration related to (b) is to pull bones out of the bone box, and to ask the class to name the bone and then to say which shape category it belongs in.

4. Applications

a. Most students are aware of the importance of dietary calcium for bone growth and maintenance but were probably unaware of the importance of the bone matrix as a storage depot for calcium homeostasis. To dramatize the significance of this store, let them know that if the parathyroid glands are surgically removed, a host of problems appear including possibly fatal cardiac *arrhythmias*, caused by the resulting *hypocalcemia*. Calcium ions are needed for many physiological processes, such as muscle contraction, nerve impulse conduction, blood clotting, and membrane transport systems. Again, this is a concrete example of the interactions among organ systems within an organism.

b. The age of a child can be estimated by examining radiographs (x-rays) of bone. The dimensions of epiphyseal plate of cartilage, extent of mineralization (bone density), size, and related features are quantified.

c. To dramatize the dynamic nature of bone, ask the students to consider the problem of astronauts in space under zero gravity where they lose a large fraction of their bone mass and strength. What kinds of interventions can they imagine that might prevent the osteoporosis of microgravity? What do national space agencies such as NASA do about the problem? How serious a problem does osteoporosis pose for long-duration space flight?

d. Andre the Giant (Andre Rene Roussimoff), world-renowned professional wrestler and actor, suffered from pituitary gigantism and acromegaly. Acromegaly develops in those for whom hypersecretion of growth hormone continues past epiphyseal closure. After closure, elongation of endochondral bones is over but appositional growth of these bones and intramembranous bones continues. This leads to deformity of the face, skull, and mandible, a coarsening of facial features, and thickening of fingers and toes.

e. To motivate a discussion about the role of diet in supporting health of the skeletal system, ask the class, "What is the absolutely best food to eat to promote bone health?" Write down their guesses. [Milk, soy, etc.] The answer is BONE! So eat foods like whole stewed fish, chicken, and ribs with bone in. The broth will extract the mineral from the skeleton and will become a rich source of both calcium and phosphate.The main protein in bone is collagen but that is made from nonessential amino acids and is thus not a big deal nutritionally. Including acidic foods such as vinegar, citrus, or tomato in the soup or stew will greatly increase mineral extraction as will long, slow cooking (crock pot is perfect).

5. Common Student Misconceptions/Problems

a. A difficult concept to visualize is that bones are constantly being broken down and rebuilt, being "remodeled." To help students understand this phenomenon, relate it to familiar processes. For example, point out that, although the skin seems to stay the same, they have learned that it is constantly undergoing a process of mitosis, differentiation, and programmed cell death. They also know that when subjected to friction or pressure, it responds by getting thicker and tougher. By analogy, this should help them understand the dynamic nature of bone tissue, not only in the service of calcium homeostasis but also in adapting to the mechanical load. Finally, point out that when a fracture heals, bone expresses the building up and breaking down processes in a dramatic fashion.

b. It is easy to forget that bones are organs. Remind the students that there are two epithelia, vascular tissue, nervous tissue, and lymphatics, in addition to the osseous connective tissue.

c. Distinguish between gigantism and acromegaly. Why do intramembranous bones continue to enlarge in acromegaly but endochondral bones only get thicker, not longer? Why does excess growth hormone lead to such different outcomes based on the age at onset?

d. *Intramembranous* does not mean inside a cellular membrane. It uses the more general meaning of *membrane*, that is, a thin, extended structure. Relate this meaning to the membranes that cover or line the body.

6. Vocabulary Aids

a. With the introduction of the term osteo*blast*, explain that the word stem *blast* often refers to an immature stage in cellular development (e.g., blastomere, erythroblast). This stem will reappear in relation to hematopoiesis. In this case, the stem refers to bone development.

b. Suggest that they associate the "b" in the stem *blast* with the word *build* as a way of remembering that osteoblasts build bone.

c. Indicate that the term *osteoprogenitor* can be understood by breaking it into parts: *osteo*—bone, *pro*—before, *genitor*—generate.

d. Stress the distinction between osteo*blasts* and osteo*clasts*. If *osteo* means bone, what do the suffixes tell you? Suggest that "blast" should remind them of *explosive growth*; "clast" should remind them of destruction (icono*clast*, someone who smites an icon).

e. Associate the word *lam*ellae with a plastic sheet used to *lam*inate.

f. The term *trabecula* literally means a "little beam" and is used to describe a network of interconnecting bridges of tissue.

g. *Hypo–calc–emia*—dissect that word with them piece by piece on the board so they see it literally means "low–calcium–in the blood."

h. Distinguish among the terms *osteogenesis*, *ossification*, and *calcification*; these often confuse students. "Calcification" is a general term for the deposition of calcium salts in soft tissue, and is not specific to bone. Calcification occurs during ossification. "Ossification" refers to the developmental deposition of mineral in bone tissue, and "osteogenesis" means bone formation and so includes other cellular and biochemical processes needed in forming "osseous tissue," that is, bone.

i. Just for the fun of it, point out the most obscure word in the chapter, diploë. Invite them to find it (Figure 6–2(b), p. 188) , and promise it will not appear on your exams!

CHAPTER 7

The Axial Skeleton

Introduction

The upright posture of human beings, enabled by the functional properties of the axial skeleton, is unique among vertebrates. Ideally, it will help your students to develop their analytical powers and scientific imagination to develop a list of what might be the adaptive advantages (and maladaptive disadvantages) of our skeletal design. Lead them in this effort using a white marker board (turn off the darn PowerPoint!). A few hints: This design relieves the forelimbs of their role in locomotion, making the hands available to evolve to manipulate objects in a dexterous way; upright posture eases carrying a big head with its big brain; the head can swivel widely, enabling a wide field of view; the orbits point forward in support of binocular vision. Also, have them look at the downside of this design: the abdominopelvic contents at risk of prolapsing through the pelvic foramen; danger of falling; herniation of intervertebral discs; heavy loading of lower appendicular skeleton (hips, knees, ankles, feet). In lab, direct their attention to articulations between the base of the cranium and the spinal column, and to articulations between vertebrae; encourage them to compare the four regions of the spinal column and compare them functionally and anatomically from this viewpoint.

Instructional Goals/Learning Outcomes

1. Identify the bones of the axial skeleton, and specify their functions.
2. Identify the bones of the cranium and face, and explain the significance of the markings on the individual bones.
3. Identify the foramina and fissures of the skull, and cite the major structures using the passageways.
4. Describe the structure and functions of the orbital complex, nasal complex, and paranasal sinuses.
5. Describe the key structural differences among the skulls of infants, children, and adults.
6. Identify and describe the curvatures of the spinal column, and indicate the function of each.
7. Identify the vertebral regions, and describe the distinctive structural and functional characteristics of vertebrae in each region.
8. Explain the significance of the articulations between the thoracic vertebrae and the ribs, and between the ribs and sternum.

Teaching Strategies

1. Lecture and Lab Ideas

a. The ridges, bumps, notches, and projections are evidence of muscle or ligament attachment sites on the bones.

b. Foramina provide places for nerves and blood vessels to pass through bones. Remind them that the nutrient foramina are established during osteogenesis.

c. Linger on Figure 7–1(a). Exactly what is included in the axial skeleton? Ask the class: What's here? What's missing? Make a list with two columns, *Included*, and *Not Included*.

d. Lay it out: There are 22 skull bones, 8 cranial and 14 facial. "We" are going to learn them all. Figure 7–3(c) nicely shows the 8 cranial bones. Take time with all the views in Figure 7–3. Many are unfamiliar. Figure 7–11 shows many of the facial bones.

e. Encourage students who might harbor "funny" feelings about looking at the human body or its parts to meet with you during regular office hours. Now is the time to get concerns of that kind identified and sympathetically dealt with. The skull for example can provoke fear of death in some.

2. Analogies

a. The ethmoid is one of the most difficult bones to visualize since it is not completely discernible in the intact skull. The views in Figure 7–3 (c) and (d), Figure 7–4 (a) and (b), and the isolated ethmoid itself in Figure 7–9 should be helpful. Describe the bone as follows. Imagine the ethmoid as a capital letter "T." (Make a "T" with your hands, as you would if signaling a time-out.) Explain that your vertical hand is analogous to the perpendicular plate of the ethmoid, which is only visible through the nasal cavity. The knuckles of the hand that form the cross bar of the "T" are analogous to the crista galli, which can only be viewed from inside the cranial cavity. The tips of your fingers and your wrist of that same hand represent the lateral masses, which can only be viewed within the orbits.

b. The fibrous cartilage pad that forms the intervertebral disc is like an "air ride" shock absorber or boat bumper: The tough outer fibrous tissue encloses a fluid cushion. Over time, the fibrous annulus can tear and permit the viscous liquid core to herniate outwards. Mistakenly called a "slipped disc," this back lesion can lead to pain and weakness. Laminectomy and spinal fusion are often applied as surgical treatments with varying success.

c. When contrasting C_1 and C_2 with the other vertebrae, mention that the *atlas* holds up the "globe" (the head), just as Atlas held up the world in Greek mythology. The manner in which the cranium's occipital condyles glide on the atlas's superior articulating surfaces allows us to nod "yes." In contrast, point out that the rotation of the atlas around the *dens* of the *axis* permits us to shake our head "no." The "globe" *rotates* on its axis.

3. Demonstrations

a. Indicate that the petrous portion of the temporal bone houses structures of the middle and inner ear. This can be better visualized if a cotton swab or a pipe cleaner is placed in the external auditory meatus and another one is placed in the internal auditory meatus.

b. Have the student palpate his/her own mastoid process. Associate the word *mastoid* with *mastos*, referring to a breast-like shape (recall *mast*ectomy, removal of the breast).

c. Using both images [for example, Figure 7–4(b)] and actual skulls, demonstrate the cribriform plate in relationship to the nasal cavity and to the olfactory bulbs. This will help clarify that the olfactory nerves (CN I) pass through its foramina.

d. From a bone box, align the vertebrae from atlas to coccyx in order on a table. In this way, the student will appreciate the manner in which the superior articulating surfaces of one vertebra align with the inferior articulating surfaces of the superior vertebra, and how their junction forms the intervertebral foramina. First, discuss and demonstrate the similarities among vertebrae, and then analyze the differences that make each region unique. Let the students practice distinguishing among representative bones from each region. Emphasize that the cervical vertebrae possess unique transverse foramina; they provide a passageway for the vertebral blood vessels that serve the brain.

e. Point out as a demonstration (with real skeletal remains, ideally) the radial tuberosity and suggest that it responds to loads from biceps brachii, which inserts there. Ask them to predict the relative size in someone accustomed to heavy lifting. Tell them that forensic pathologists look at such features when trying to profile an individual from skeletal remains. Also, note how the roughness of the surface provides more sites for tendon attachments.

f. Use human and animal skulls (dog, deer) cut to reveal the area where the sense of smell enters the brain (cribriform plate). It's illuminating to compare their relative sizes in two creatures with very different olfactory capabilities.

4. Applications

a. If the perpendicular plate of the ethmoid bone is displaced from its medial position narrowing the nasal cavity (deviated septum), the result can be chronic sinus problems.

b. The lamina is the portion of the vertebra that is removed during a laminectomy. Laminectomies are performed most commonly to relieve spinal stenosis, a narrowing of the vertebral canal or to treat disc herniation.

c. The xiphoid process is used as a palpable landmark during CPR administration.

d. The vertebral regions (cervical, thoracic, etc.) are applied in clinical neuroanatomy to refer to spinal nerves and spinal segments.

5. Common Student Misconceptions/Problems

a. Students often lose the perspective that the cranial bones have an inside and outside surface. For example, externally, the frontal bone forms the forehead, and internally forms the roof of the orbits. The *Focus—The Individual Bones of the Skull* (Figures 7–5 to 7–12) in the text beautifully demonstrates both external and internal views of the bones, and should solve this problem. Recommend that they spend adequate time studying and understanding the anatomy taught in that set of figures.

b. It is easy to confuse the jugular and carotid foramina. In labs where actual skulls or accurate plastic casts are available for study, view the cranial cavity internally.

The petrous process of the temporal bone projects between the two openings; the carotid foramen is anterior to the petrous process. Viewing externally, the carotid foramen is still anterior, but because the carotid canal travels laterally and posteriorly, the two foramina sit next to each other, with the styloid process situated laterally.

c. Many students think of the maxilla simply as the bone that holds the upper teeth without realizing that it composes much of the face, including the floor of the orbit and the roof of the mouth. Carefully trace the outline of the maxilla, relying ideally on a real skull, assisted by Figures 7–3 (a) and (b), and 7–11.

d. Some students may expect that males have one fewer rib than females. Point out that, in fact, humans of either gender most often have an even number of ribs, and that an odd number is no more common an anatomical anomaly in one gender or the other.

e. To a layperson, a skull is commonly seen as a sign of death. To an anatomy student, it's a beautiful, exquisitely complex and intricate assembly of 22 bones that house, support, and protect the human mind and mouth!

f. The skull includes *both* the cranium and the facial bones.

6. Vocabulary Aids

a. The term *ethmoid* derives from "sieve like," and refers to the cribriform foramina and numerous ethmoid sinuses of the ethmoid bone. *Crib*riform may remind students of a corn*crib*.

b. Associate the Latin adjective *magnum* with the word *magnify* to indicate that this foramen is BIG. Most students know what it means already. Ask the class, "How many know what the word magnum means?" (Note, *not* who knows . . .)

c. The parietal bones get their name because they form the *wall* of the cranium, just as a parietal membrane covers the *wall* of a body cavity.

d. The squamous portion of a bone refers to the flattened region, just as squamous epithelium refers to flattened cells.

e. Anatomically, a *pro*cess is a bone feature that "sticks out" or *pro*jects.

f. Associate the word *petrous* with the word *petrified* to emphasize the hardness of the petrous portion of the temporal bone, which suits it to protecting the delicate inner ear. It looks like a rock outcropping.

g. Indicate that "styloid" derives from *stylus*—a long, pointed object used to mark in clay.

h. *Sphenoid* means "wedge-shaped." The sphenoid bone is "wedged" between the cranial and facial bones. Trace the outline of the "bat with wings" that the sphenoid resembles. This form is seen very clearly in Figures 7–4(b) and 7–8.

i. Hypophysis, meaning "to grow beneath (the brain)," is the preferred term for the pituitary gland.

j. The pituitary gland "rides" protected in its sella turcica, literally, "Turkish saddle."

k. *Ptery*goid means wing-shaped. The *ptero*dactyls were winged reptiles that lived until the late Cretaceous era.

l. Coronoid comes from the word *corona*, a crown. Imagine this process as the point in a king's crown. The word *coronation* is related.

m. The word *mental* pertains to the chin. Recall Rodin's *The Thinker*.

n. The hyoid bone "hides" behind the mandible.

o. The coccyx gets its name because this bone was thought to resemble the beak of the cuckoo bird (*coccyx* means cuckoo in Greek). Anticipate the confusion between coccyx and coxal bone that Chapter 8 will generate.

p. The cranium is also known by the very apt term, *braincase.*

q. The *vomer* is probably the least famous skull bone. It means *plowshare* (from its shape).

CHAPTER 8

The Appendicular Skeleton

Introduction

Introduce this topic by continuing the thread begun in the last chapter on bipedal or upright locomotion. In glaring departure from the standard mammalian body plan stands the body of man, nowhere seen more clearly than in the isolated skeletal system as in Figure 8–1. In this, the second chapter to deal in detail with the human skeleton, the two girdles and four limbs are studied. The wide pelvic girdle, the stout bones of the leg, the unique locking knee joint, and the human foot are all striking adaptations to facilitate bipedal locomotion. They complement the heavy lumbar vertebrae and compound curvature of the spine that was described in Chapter 7. Just as support and visceral protection resides in our axial skeleton, so does our ability to walk about, to kneel, to grasp and manipulate, to caress, and to embrace reside in our appendicular skeleton. Humans have evolved a precision in the use of their hands that is unrivaled by any other creature, a benefit made possible by no longer requiring their use in locomotion. You might begin by asking the students how a paleontologist might tell if a fossil bone was human or primate. Use this question to bring out the skeletal differences outlined above. A good introduction to these matters can be found in Roger Lewin's *The Origin of Modern Humans* (W.H. Freeman & Co., 1998). A valuable online resource is the PBS Evolution Library (http://www.pbs.org/wgbh/evolution/library/).

Instructional Goals/Learning Outcomes

1. Identify the bones that form the pectoral girdle, their functions, and their superficial features.
2. Identify the bones of the upper limbs, their functions, and their superficial features.
3. Identify the bones that form the pelvic girdle, their functions, and their superficial features.
4. Identify the bones of the lower limbs, their functions, and their superficial features.
5. Summarize sex differences and age-related changes in the human skeleton.

Teaching Strategies

1. Lecture and Lab Ideas

a. During the laboratory study of the appendicular skeleton, it will benefit the students if they adopt a systematic approach to study each bone:

   1. Examine the bone, paying attention to general shape and gross anatomy. Ideally, real human bones in a bone box or as an articulated skeleton will be available. Failing that, high-quality resin casts of real bones can be substituted. If these are not available, use the textbook figures. Avoid cheap plastic imitation bones.

   2. Sketch the bone and label the detailed characteristics.

   3. Hold the bone up to an articulated skeleton or an anatomical chart to compare shape and orientation. This procedure aids in the visualization of the bone *in situ*. During this comparison, have them find conspicuous landmarks that will let them decide if the bone is left or right. Two landmarks are required, one to distinguish medial from lateral, and the other anterior from posterior. Have them use these observations to reconstruct on a lab table an anatomically correct human skeleton. This works particularly well as a collaborative, consultative group activity. Give the opportunity to "present" the reconstructed skeleton bone by bone to the rest of the lab group.

b. As each bone is examined, note the landmarks that can be used to determine medial from lateral (m/l) and anterior from posterior (a/p).

   1. **Clavicle:**

      (m/l): The sternal (medial) end is flat, almost as if it were sawed off.

      (a/p): When the sternal end is in position, the S shape of the bone curves away from the body and then into the body to meet the acromion process.

   2. **Scapula:**

      (m/l): The glenoid fossa is lateral.

      (a/p): The spine is posterior.

   3. **Humerus:**

      (m/l): The round head and the more prominent medial epicondyle are medial.

      (a/p): The deep coronoid fossa is posterior.

   4. **Ulna:**

      (m/l): The styloid process is located at the medial side of the wrist.

      (a/p): The olecranon process is posterior.

   5. **Radius:**

      (m/l): The styloid process is located at the lateral side of the wrist.

      (a/p): The anterior surface of the distal extremity is smooth as compared to the posterior side.

6. **Coxa:**

   (m/l): The acetabulum is lateral.

   (a/p): The sacroiliac joint is posterior.

7. **Femur:**

   (m/l): The large, round head is medial.

   (a/p): The intercondylar fossa is posterior.

8. **Patella:**

   (m/l): The larger of the two facets is the lateral facet.

   (a/p): The anterior surface is convex.

9. **Tibia:**

   (m/l): The large malleolus is medial.

   (a/p): The tibial tuberosity is anterior.

10. **Fibula:**

    (m/l): The articular facets are medial.

    (a/p): The shallower, slightly superior facet is anterior.

c. When discussing the humerus, point out that the tendon from one of the heads of the biceps brachii travels along the intertubercular groove.

d. The disc-shaped radial head is unique; it allows for a rotational movement between the radial head and the capitulum.

e. As the largest bone of the body, the femur is relatively easy for students to identify. Its robustness relates to its weight-bearing role. Point out the greater trochanter, linea aspera, and gluteal tuberosity as points of massive muscle attachment.

f. Mention the mnemonics from the text (pp. 253, 261) for the eight carpal bones, "Sam Likes To Push The Toy Car Hard" and the seven tarsal bones, "Tom Can Control Not Much In Life."

g. The calcaneus is actually the fusion of two bones; this is why there are only seven tarsals in contrast to eight carpals.

h. When showing Figure 8–2(a), indicate that the shoulder girdle is merely 2 bones/side, the clavicle and the scapula. The only attachment of the shoulder girdle to the axial skeleton is the articulation between the sternal facet of the clavicle to the manubrium at the claviculosternal joint. When projecting Figure 8–3, give the class adequate time to simply appreciate the graceful beauty and subtle purposes of the scapula.

i. Be sure, using Figure 8–10, to characterize the differences between the male and female pelvis. This is the sort of information they may encounter in CSI or another program that involves forensic anatomy.

2. Analogies

a. The small, rounded surface of the capitulum of the humerus is reminiscent of the top of someone's head, where a "cap" would be placed.

b. The coronoid process of the ulna resembles a point in a king's crown, just like the coronoid process of the mandible. Likewise, the styloid process has the same shape as the styloid process of the temporal bone in the skull, both named for *stylus*, a sharp pointer.

c. The acetabulum is like a pie cut into three pieces: the ischium, ilium, and pubis all contact within it.

3. Demonstrations

a. Emphasize the structural significance of the coronoid, olecranon, and radial fossae of the distal humerus by demonstrating how the processes articulate into the respective fossae when the forearm is flexed and extended. An articulated skeleton or three-dimensional model is invaluable here. Online animations exist as well.

b. There is a tendency to confuse the ulna and the radius. If you rotate your wrist, the forearm bone that rotates (i.e., moves in a circular path) is the *radius*. Tell them the knob we feel at the *el*bow is the o***l***ecranon of the u***l***na. Alternatively, suggest they call the joint the ULbow!

c. With reference to their own or another's hands, have them note the number, proportion, and absolute length of the phalangeal bones of each finger. This makes a good introductory lab exercise. Students learn to apply the metric scale, keep records of actual data, and struggle with the uncertainty of deciding where the "edge" of a phalanx lies.

d. Instruct the students to run their fingers along their "shin bone" to illustrate the location of the anterior margin of the tibia.

e. The calcaneal tendon is made easily visible if the foot is dorsiflexed. This tendon is often called the "Achilles tendon," after the Greek mythic figure of Homer's *Iliad* whose mother held him by his heel as she dipped him into the River Styx to make him invincible. An arrow to his "Achilles heel" mortally wounded him at the battle of Troy.

f. To demonstrate that the upper limb is designed for maximum flexibility, perform a wavelike undulation of your whole upper limb in front of the class. Invite them to join in, briefly. This extraordinary range and form of motion are the rewards we get for living with chronically sore feet, the common lot of human kind!

4. Applications

a. The only place the scapula is attached to another bone is to the lateral end of the clavicle; thus, an individual who has a fracture of the clavicle will often have the appearance of their shoulder "drooped" as compared to the intact shoulder.

b. The clavicle is the most commonly fractured bone because it is the only connection from the shoulder to the sternum. The thin neck where the two curves meet is the likeliest to break.

c. Use a study of Figure 8–13 to illustrate how a sudden outward and backward displacement of the foot might fracture both the lateral and medial malleoli, the *Pott fracture*, a common athletic injury. (*Note*, eponyms usually lose their possessive case in this edition, consonant with modern usage.)

5. Common Student Misconceptions/Problems

a. Although students usually have no difficulty identifying the scapula as the posterior shoulder blade, they often have difficulty envisioning that the coracoid process hooks around anteriorly to contribute to forming the glenoid fossa. This can be clarified by referring to an articulated skeleton and by pointing out that the coracoid process is one of the origins of biceps brachii, a plainly anterior muscle.

b. Students like to associate the humerus with their "funny bone." While this may be an effective way of remembering the humerus, remind them that it is *not* spelled "humorous!"

c. Help the students to distinguish the *anatomical* and *surgical* necks of the humerus. The anatomical neck lies between the head and the tubercles. The surgical neck (so named because it is a common site of fracture) is the metaphysis of the developing bone.

d. Students easily confuse the names for the two bones of the lower leg ("tibula" and "fibia"). No easy mnemonic here, but it is helpful to warn them about the possible confusion and encourage them to practice the distinction. Tibia is the stout one; fibula is the diminutive one.

e. Anatomically, "arm" means upper arm. Upper limb = arm + forearm + wrist + hand.

6. Vocabulary Aids

a. Girdles (fortunately) are no longer familiar garments. Point out that *girdle* comes from the word *gird*, which means to enclose or wrap around. Hence, the pectoral girdle refers to the bones that enclose the pectus (chest), and the pelvic girdle cradles the pelvic cavity.

b. The word *clavicle* comes from the same root as the *key*board instruments, the clavichord and clavier; clavicle derives from Latin for "little key."

c. To clarify which is the *coracoid* and which is the *acromial* process of the scapula, point out that the coracoid process hooks around anteriorly (coracoid means crow-like). *Acro*mial suggests *acro*bat or *acro*polis; the acromial process is the *highest* point of the shoulder just as acrobats may perform high on the trapeze and acropolis is the city on the hill.

d. Tell the students that the word *trochlea* means "pulley." The trochlea of the humerus resembles a small pulley or spool that the trochlear notch of the ulna rotates within during forearm flexion.

e. The word *metacarpal* literally means "beyond the wrist." Ask them what the word *metatarsal* means.

f. The word *phalanges* comes from *phalanx*, "a row of soldiers."

g. A common misspelling of ilium is "ileum." Give this as an example of how an error in a single letter could lead to confusion in a patient's chart.

h. The word *acetabulum* literally means "vinegar (acetic acid) cup."

i. The word *malleolus* means "little hammer." These two bony projections above the ankle look like the head of a hammer.

j. The **t**alus is on **t**op and articulates with the **t**ibia.

k. The navicular is boat-shaped. Its name comes from the same root as *nav*al and *nav*igate.

l. The word *cuneiform* means wedge-shaped.

m. The textbook abandons *os coxa.* It uses the English *coxal bone* instead.

CHAPTER 9

Articulations

Introduction

The potentially disabling joint disease **rheumatoid arthritis** (RA) and the new strategies for the care of RA patients that have developed by applying basic biomedical knowledge (implants, monoclonal antibodies, immune modulators, artificial synovial fluid, engineered tissue grafts) can be used to motivate interest in the anatomy and physiology of joints. RA is an autoimmune illness, that is, a disease caused by an attack by the body's immune system upon one's own cells *and* extracellular matrix. It affects between 0.5 and 1.0 percent of the adult population. RA commonly targets synovial joints leading ultimately to severe inflammation and crippling destruction of the joints. Students need to distinguish between RA and *osteoarthritis*, the wear-and-tear arthritis that often appears with advancing years. One hypothesis is that RA (and other autoimmune diseases) result from a bacterial or viral infection that tricks the immune system to attack and destroy articular *collagen* (and in turn, the articular cartilage) because the articular form of collagen shares an antigenic determinant with the pathogen. If you are searching for a good research topic to offer your students, how about looking into the basic science aspects of RA?

Instructional Goals/Learning Outcomes

1. Contrast the major categories of joints, and explain the relationship between structure and function for each category.
2. Describe the basic structure of a synovial joint, and describe common synovial joint accessory structures and their functions.
3. Describe how the anatomical and functional properties of synovial joints permit dynamic movements of the skeleton.
4. Describe the articulations between the vertebrae of the vertebral column.
5. Describe the structure and function of the shoulder joint and the elbow joint.
6. Describe the structure and function of the hip joint and the knee joint.
7. Describe the effects of aging on articulations, and discuss the most common age-related clinical problems for articulations.
8. Explain the functional relationships between the skeletal system and other body systems.

Teaching Strategies

1. Lecture and Lab Ideas

a. Many students, when prompted to think about articulations, will likely picture the knee or some other prominent synovial joint. Correct this limited view when you begin the topic by clarifying that an *articulation* means simply *any* joining of two bones, and doesn't require that they move. You can (literally) point to bones of the skull as familiar examples of immovable joints. From their studies of the cranium, they already know about a synarthrosis without knowing the term. They will soon be burdened with many obscure terms by which joints may be characterized and classified.

b. An easy mistake to make in presenting this topic is to begin with multiple classification schemes based on arcane terms whose meaning or significance is not obvious. This will overwhelm most students and become a barrier to learning. Better to begin with a simple three-category functional scheme: First, list the English descriptions ("no movement" etc.), and then add the technical descriptors (synarthrosis, etc.). Develop a table in a conversational style on the board, element by element. That is, don't drop Tables 9–1 and 9–2 on them all at once but project them and review them as a summary of what you have already covered with marker by hand.

c. Figure 9–6 offers a nice way to illustrate the many classes of synovial joints. The gross anatomy, structural models, and actual examples merit systematic review in class. Point out as you do that even "freely movable" synovial joints differ greatly in their ease and range of movement.

d. The principle that there is a trade-off between stability and mobility can be clearly illustrated using Figure 9–9 of the shoulder joint. As you review the gross anatomy, point out how much of the anchoring of the joint depends on relatively weak connective tissue (rotator cuff tendons, ligaments, glenoid labrum). Also, the glenoid cavity is rather shallow. Compare the shoulder to the elbow (Figure 9–10), a very tight-fitting, uniaxial humero-ulnar joint. This concept then leads naturally to a presentation of the robustly trussed (but heavily loaded) hip (Figure 9–11) and knee (Figure 9–12) joints. Figure 9–12 (c) and (d) offer rare views of freshly dissected human bones and joints.

2. Analogies

a. Describe flexion as encompassing those movements that result in a fetal position.

b. Bursae can be compared to boat bumpers or dunnage bags used in freight cars and trucks to cushion objects.

c. The humero-ulnar (elbow), the femoral-tibial (knee), and tibial-talar (ankle) joints can be analogized to door hinges, a familiar uniaxial "joint."

d. The Delta-style faucet is an analogue of ball-and-socket joints such as the shoulder and hip. The type of faucet even incorporates an equivalent to a labrum made of plastic!

e. A dovetail joint in an antique wooden box or drawer is a "synarthrosis" held together with collagen-based (hide) glue.

f. The femoral condyles look like big knuckles upon which the femur rests its weight.

3. Demonstrations

a. Unlike synarthroses and amphiarthroses, diarthroses have a cavity between the bones surrounded by a capsule. This cavity is lined with a synovial membrane, and filled with synovial fluid. The cavity and capsule give this joint its freedom of movement. A simplified example of such a structure can be demonstrated using two dowel rods held together with a piece of rubber tubing. The sticks are the bones, the tubing is the capsule, and the space inside the tubing is the synovial cavity.

b. As a lab activity, have teams of students demonstrate pairs of movements (extension–flexion, abduction–adduction, etc.). As they "perform," get them to lead the other students and say aloud the names of the movements as they are done. Have them rely on the series of photos in the text (Figures 9–3 to 9–5).

c. The use of fully articulated three-dimensional models of representative joints (obtainable from many lab supply houses) can be a very effective way to bring out the principles of diarthroses.

d. Artificial synovial fluid (e.g., Synvisc® made from rooster combs) used to treat osteoarthritis might be available from a cooperating orthopedist or drug company representative—it is beastly expensive. Alternatively, a solution of hyaluronic acid, available from biochemical supply houses such as Sigma, will demonstrate a viscoelastic fluid. Permit students to manually examine the material to gain a clear idea of a biological viscoelastic lubricant, a springy liquid! Fresh egg white will do in a pinch.

e. Ask the class to point at the location of the only bone in their body that doesn't articulate with another bone. They should be pointing at the space between the muscular base of the tongue and the soft tissue above the thyroid cartilage (hyoid bone).

4. Applications

a. A whiplash injury to the neck is the result of *hyper*extension.

b. Full body flexion ("fetal position") is recommended for lumbar puncture ("spinal tap").

c. In physical rehabilitation after joint injury or surgery, quantitative measurements of the angular range of motion (ROM) are made as a guide to progress.

d. Man-made synovial fluid (Synvisc®) can be injected within the synovial joint capsule in tiny amounts to treat the pain and disability of osteoarthritis.

e. To reinforce the concept that mobility is purchased at the price of instability point to the fact that the shoulder joint is the one most commonly dislocated.

5. Common Student Misconceptions/Problems

a. To most naïve students, "joint" means a visible feature such as the knee; most don't visualize the interior structure nor possess the basic concept, namely the joining of *two bones.* Thus, it is fundamentally important to establish the concept that an *articulation* is simply the joining of two bones, irrespective of the design or movement.

b. Students may have trouble sorting joints into three functional categories. To overcome this, first remind them that they are now familiar with the concept of categorization (they have seen organ systems, epithelia, and connective tissues sorted into categories).

Tell them to ask these three questions about each joint: How freely does it move? What holds it together? Is three a synovial capsule? Ideally, as they examine the skeleton, they will see that each joint belongs in one of three functional categories. Moreover, they will learn to apply structural specifics to further subdivide the articulations.

c. Certain types of joint movements may be difficult for students to comprehend. For example, "flexion" and "extension" are easy to distinguish when applied to uniaxial joints but become cryptic for a multiaxial joint. The simple model of articular motion (Figure 9–2) using a pencil on paper is a useful tool for visualizing movements. Also, performing and naming the movements by pairs will make the learning task more concrete and therefore more manageable.

d. The anterior and posterior cruciate ligaments are named from their *points of attachment* on the *tibia*. They cross the joint within the knee capsule and attach to posterior and anterior positions on the femur, respectively. Because they angle, the cruciates provide stability in both lateral–medial and anterior–posterior directions.

6. Vocabulary Aids

a. For a suture, compare the uniting of two skull bones with fibrous connective tissue to the sewing together of two pieces of fabric with a zigzag stitch. The word *suture* comes from *sutura*, Latin for seam.

b. Dissect the word *syn–chondro–sis* to illustrate that it literally means "[held] together with cartilage." Point out that a synchondrosis can become a *synostosis* if ossification of the cartilage occurs.

c. The word *cruciate* means "crossed"; *crux* is Latin for cross. *Crucify* derives from that root.

d. When someone is "*abd*ucted," they are taken *away*, just as abduction takes the limb *away* from the body. During "*add*uction," the limb is "added" to the body.

e. When someone is prone, they are laying face-down. Imagine the palm as the "facial" side of your hand. Pronation of the hand is to position the palm downward or posteriorly. When your hand is supinated, it is in the position of being ready to hold a bowl of soup ("*soup*inated").

f. The word *synovial* means "like egg white," that is, a clear, slippery, viscoelastic fluid.

g. Point out that the interspinous and supraspinous ligaments get their names because they are attached to the *spinous* processes of the vertebrae, not because they stabilize what laypersons call the "spine."

h. The *dens* of the axis looks like a *dunce* cap. [*Dens* is the Latin word for tooth.]

CHAPTER 10

Muscle Tissue

Introduction

Students are often interested in the relationship of athletic training to athletic performance. This interest focuses attention on the ultimate source of muscle strength, the myofibril. How does exercise affect myofibrils? Some forms trigger the process of hypertrophy (bulking up). Not all exercise causes hypertrophy. Comparing sprinters to marathon runners makes this apparent. What types of exercise cause hypertrophy? What are the consequences of endurance training? What is the cellular mechanism of hypertrophy? What are the cellular adaptations for endurance? What is the role of nutrition in building muscle mass? A related area of interest is muscle cramps, contractures, and post-exercise muscle soreness. Many of these issues are poorly understood; it is useful to point to such questions, especially familiar ones, where our knowledge is spotty. This gives the students a sense that anatomy and physiology are active sciences and not static collections of dull facts.

Instructional Goals/Learning Outcomes

1. Specify the functions of skeletal muscle tissue.
2. Describe the organization of muscle at the tissue level.
3. Explain the characteristics of skeletal muscle fibers, and identify the structural components of a sarcomere.
4. Identify the components of the neuromuscular junction, and summarize the events involved in the neural control of skeletal muscles.
5. Describe the mechanism responsible for tension production in a muscle fiber, and compare the different types of muscle contraction.
6. Describe the mechanisms by which muscle fibers obtain the energy to power contractions.
7. Relate the types of muscle fibers to muscle performance, and distinguish between aerobic and anaerobic endurance.
8. Identify the structural and functional differences between skeletal muscle fibers and cardiac muscle cells.
9. Identify the structural and functional differences between skeletal muscle fibers and smooth muscle cells, and discuss the roles of smooth muscle tissue in systems throughout the body.

Teaching Strategies

1. Lecture and Lab Ideas

a. Remind the students that muscle tissue, which they met briefly amidst the storm of tissue types presented in Chapter 4, can be sorted into three distinct types, based on how the contractile machinery is organized and how the nervous system figures in its control. Make a simple table on the board summarizing this basic framework: (1) skeletal muscle (striated, voluntary-somatic); (2) smooth muscle (nonstriated, involuntary-visceral); (3) cardiac muscle (striated, involuntary-visceral). Tell them that the muscle type about which we know the most is skeletal muscle, the kind that composes the muscular system, and that it will be the subject of most of your presentation.

b. When taking the class through the many panels in Figure 10–1, take pains to be sure students understand that (1) some views are longitudinal and others are cross sections, and (2) views move to finer and finer structural detail. These may seem obvious to you as the teacher, but they often elude students. Since this large-to-small pedagogic strategy is often encountered in the text, it is worth belaboring now.

c. Keep the alphabet soup to a bare minimum: "A band," "I band," and "Z line" is plenty. Don't permit nomenclature to obscure the main facts of contraction: myosin heads sticking off the thick filaments use ATP to pull the actin filaments toward the middle of the A band, shortening the sarcomeres and thus pulling on an external load.

d. The sequence of structures and processes that lead from motor neuron impulse to the twitch of its muscle fibers is one of the best understood in human biology. It also represents an excellent opportunity to help your students develop the intellectual skills and habits of *causal reasoning* that underlies science and medicine. A very useful tool in this effort is the Navigator "An Overview of Skeletal Muscle Contraction" (Figure 10–9). Students will find that once the logic of a causal chain or sequence is understood, learning and remembering it will become far easier. Also, as their training progresses, ideally they will learn to apply these tools to understanding the diagnosis and treatment of illness.

e. Not only is neuromuscular transmission crucial to movement, it also serves as a model for understanding synaptic transmission between neurons. Applying the causal reasoning approach mentioned above will facilitate learning this vital information. Take a step-by-step approach to the operation of the NMJ, helping students to understand what is *presynaptic* and what is *postsynaptic*. Figure 10–9 shows where it fits in the overall process, and Figure 10–10 details the structures and steps of neuromuscular transmission. As before, alert the students to the change of scale between panel (b) and the Step 1 panel. Encourage them to use their powers of visualization to imagine unseen events at the molecular level.

f. Compare aerobic vs. anaerobic metabolism. The former produces more ATP but requires O_2; the latter doesn't require O_2 but produces less ATP. How is this difference reflected in how aerobic and anaerobic motor units are used?

g. When comparing the properties of fast-red, fast-white, and slow motor units, key the distinction to fatigue resistance. That is, fast-white motor units have few capillaries, a fast myosin ATPase, low glycogen reserves, few mitochondria, and no myoglobin, and rely on glycolytic metabolism; it follows that they would fatigue quickly. On the contrary, slow motor units have more capillaries, possess a slow myosin ATPase, have numerous mitochondria and myoglobin, and go through aerobic metabolism; this makes them the most resistant to fatigue. Fast-red motor units, the most abundant type, are intermediate in properties and fatigability between these two.

h. Stress that three important differences between cardiac and skeletal muscle are: (1) cardiac muscle has its own intrinsic activation system, pacemaker tissue; that is, unlike skeletal muscle, cardiac muscle does not require neural activation but is self-exciting; (2) the strength and rate of the heartbeat are controlled by the autonomic nervous system; and (3) cardiac muscle cells are short, branched, and linked together as a functional syncytium at the intercalated discs.

i. Similarly, point out the most important differences between smooth and skeletal muscle are that (1) smooth muscle is not attached to the skeleton, but occurs primarily in the walls of hollow organs; and (2) its contractions are stimulated and controlled involuntarily by the autonomic nervous system (but sometimes triggered or modified by higher centers).

j. Point out that control of contraction of smooth muscle is *myosin* based, unlike skeletal and cardiac muscles that are *actin* based. In smooth muscle, calcium ions bind to *calmodulin*, a soluble protein and not to *troponin*, a structural protein. Calmodulin activates a protein kinase that phosphorylates the regulatory light chain on myosin, permitting its attachment to actin, thus setting the cross-bridge cycle in motion. The slowness of this enzymatic system is part of the reason why smooth muscle contraction is markedly slower than striated muscle.

2. Analogies

a. Compare the terminal cisternae of the SR to holding tanks for calcium ions; when the valves (calcium channels) are opened, the calcium ions flood out and rapidly bind to troponin.

b. The sarcoplasmic reticulum fits like the sleeve of a lacy shirt over the arm (myofibril).

c. Compare F-actin to two strands of pearls that have been twisted around each other. Each pearl is a G-actin subunit. On each G-actin there is a sticky spot for myosin; at rest, tropomyosin covers the spot, preventing myosin attachment. The troponin molecules are like little men, each sitting on the twisted strand of pearls, with his legs (the actin-binding subunit) wrapped around a single pearl. He has one hand (the tropomyosin-binding subunit) holding a tropomyosin and the other hand (calcium-binding subunit) open to receive calcium ions. When calcium ions stick to his open hand, he tilts and rolls the tropomyosin away from the sticky spot. Myosin heads now attach to actin. When the calcium ions are removed, the little man returns to the resting situation sending tropomyosin rolling back to the rest position covering the myosin-binding sites on actin.

d. An analogy for a motor unit is an old-fashioned telephone party line. A single circuit (motor unit) in the telephone exchange (the CNS) is activated. This causes all the telephones on the party line (the innervated muscle fibers) to ring (contract) at the same time.

e. An analogy to the action of cross-bridges moving actin is rowing: Imagine people sitting in a canoe (the myosin) and having to glide the canoe across the surface of the water (the actin) by reaching out with their oar (cocking), putting the oar into the water (attachment), pulling the oar causing water to be pulled along and projected toward the back of the boat (pivoting), lifting the oar out of the water (detachment), swinging the oar around (activation), and reaching out again (cocking).

f. An analogy for the length–tension relationship is a tug-of-war. The side with the most people (cross-bridges) pulling will pull the hardest. As overlap between thick and thin filaments is reduced with increased sarcomere length, fewer people will be pulling and the force will go down. This analogy exposes the fundamental weakness of the sliding filament model, its inherent instability. If a thick filament moves, by chance, toward the Z line it will become stronger on that side due to increased numbers of cross-bridges. This is a positive feedback situation in which the thick filament would be pulled increasingly toward the Z line. However, this doesn't occur appreciably because the elastic connecting *titin* filament becomes stretched and so imposes a restoring force, stabilizing the system.

g. The concepts of *isotonic* (same tension) and *isometric* (same length) can be illustrated by imagining a power lifter trying to lift a 300-lb weight. He grasps the bar and grunts as he gradually increases the tension, or pull, in his muscles. When the tension just exceeds the 300-lb weight, the weight will rise and he will keep the tension constant as he shortens the muscle (isotonic contraction). Now imagine the same power lifter trying to lift a 1000-lb weight. He grunts as he gradually increases the tension in his muscles, but no matter how hard he tries, he cannot get the tension in his muscles to exceed the 1000-lb load. The tension in his muscles might be increasing, but his muscles are staying the same length (isometric contraction).

3. Demonstrations

a. Produce a simple picture of the sliding filament theory in this way: hold your hands, palms facing you, with your fingertips just interlaced and your thumbs stuck up at right angles. Tell the class that your fingers represent thick and thin filaments and your thumbs, the Z lines. Slide your fingers across each other to represent the sliding of the filaments. Point out that your thumbs come closer together, but the length of your fingers stays the same. Hence, you were able to decrease sarcomere length without changing filament length.

b. The structural arrangement of the thin and thick filaments in smooth muscle can be compared to the plastic webbing used in grocery stores to bag oranges. The strands of the webbing represent the filaments, and the cross connections represent the dense bodies. The result is a diamond pattern. When pulled in one direction, the diamonds get long and thin (smooth muscle in its relaxed state). When pulled in the opposite direction, the diamonds get short and fat (smooth muscle in its contracted state).

c. Warn the students that it is easy to get confused by muscle terminology. Learn it as a size-down sequence: fascicle (bundle), fiber (cell), myofibril (organelle), and filament (supramolecular structure). Have them sketch and assign sizes to the diameters of the structures: 1–2 mm, 50–100 μm, 1–2 μm, and 5–20 nm. This is also a good practical exercise on metric units.

d. Repeat the famous Helmholtz demonstration of the nature of unfused tetanus with your class in lecture or lab. [Hermann von Helmholtz was a major figure in 19th century physics, biophysics, and medicine.] Have the students make a fist and extend their index finger. Place the finger in the external auditory meatus firmly and instruct them to powerfully contract their forearm muscles and to listen carefully. They should hear a low frequency rumble (about 16 Hz). That sound results from the variation of tension as the motor units contract asynchronously at sub-tetanic rates! Using the ear as a force transducer! Afterwards, I always enjoy telling them "This morning you didn't know you would be putting your finger in your ear and listening to your forearm but I did."

4. Applications

a. Rigor mortis is an interesting application of the biochemistry of the cross-bridge cycle. Using Figure 10–12, ask the students where the cycle would stop if ATP were absent [Step 3 – angled and attached]. Point out that after death, ATP becomes depleted. Many will have an "a-ha" reaction when you ask if they now understand the *why* behind the "stiffness of death"? The rigor mortis eventually resolves as internal proteases break up the myofibrils at the Z lines. This endogenous proteolysis plays a major role in meat tenderness that improves as meat is "aged."

b. Skeletal muscle cells express different molecular forms of troponin subunits than do cardiac muscle cells. This lays the basis for a blood test to characterize the troponin isoforms as a way of confirming and following the progress of a heart attack (myocardial infarction) as distinct from skeletal muscle injury.

c. Since muscle contraction is a chemical reaction between actin, myosin, and ATP, lowering the temperature can slow it. This is part of the rationale for using ice packs to limit the injury in muscle cramp or strain.

d. Disuse atrophy can be made palpable by asking for students who wore a cast for a broken bone. At first the cast fits snugly, but as time goes by, the cast becomes looser as limb muscles waste from disuse. Now a coat hanger can reach the itch!

5. Common Student Misconceptions/Problems

a. A danger in presenting *three* levels of connective tissue (Figure 10–1) is that it can distract from interest in muscle fibers. Repeatedly emphasize that it is the cells (fibers) that do the active pulling.

b. The commonsense meaning of "contraction" needs to be carefully contrasted with the specialized meaning employed by muscle physiologists. It means more than "shortening"; it signifies the active state of cross-bridge cycling, irrespective of whether muscle length is decreasing, increasing, or remaining constant.

c. Students may have difficulty with the concept that a resting myosin head possesses stored energy. Point out that the ATPase activity in the myosin head splits phosphate from the ATP, releasing chemical energy. The energy is not lost, but stored by distorting the myosin head, much like setting a mousetrap or cocking a pistol. When the trigger is touched (contact with actin is made) the cross-bridge snaps to a new shape pulling the thin filament along by a tiny step.

6. Vocabulary Aids

a. The Greek combining form *my(o)* always refers to muscle. It comes from "mouse."

b. *Sarco-* is also Greek; it means "flesh." A *sarco*phagus (burial box) literally means "flesh eater."

c. Muscle cells are called "fibers" because they are long, threadlike cells.

d. Associate the "tin" in *actin* with the word *thin*. Thus, actin forms the backbone of the thin filament. Since they are th*I*n, they look l*I*ght when viewed in the microscope; hence, they form the *I* band. The *A* bands are d*A*rk. (In fact, *I* refers to optically *I*sotropic, *A* to *A*nisotropic, based on analysis by polarization microscopy.)

e. The sarcoplasmic reticulum is a specialized form of smooth endoplasmic reticulum. It forms a lacy network around the myofibril.

f. A *cistern* is a storage tank for water. A terminal *cistern a* is a storage tank for calcium.

CHAPTER 11

The Muscular System

Introduction

Learning *all* the muscles of the body is beyond the scope of most undergraduate A&P courses (and the conscientious teacher lets their class know explicitly what their expectations are). Nonetheless, many *can* be learned, especially if they are learned in terms of the movements they participate in. Another learning aid is to provide the students with some understanding of how muscles get their names, so that the task of recall is elevated above mere rote. Here's an approach that encourages collaborative learning and improved oral communication: Have the students working as a group in class generate a list of body movements (walking, lifting, facial expressions, bending, etc.). Break the class into small groups and assign a movement type to each. Using their textbook *Martini's, Atlas of the Human Body* and a lab manual as resources, have each group research the muscles involved in their assigned movements. Each group should present a description of the muscles involved, their location, origin, insertion, action, and role in the motion (agonist, synergist, antagonist). They could include the involved articulation if you wish. Encourage them to also discuss the meaning of muscle names. This activity can be spread out over several weeks, and intermixed with other activities (dissection, study of models, histology, human or animal muscle physiology). A written report might be required as well, especially if your institution "encourages" writing "across the curriculum."

Instructional Goals/Learning Outcomes

1. Describe the arrangement of fascicles in the various types of muscles, and explain the resulting functional differences.
2. Describe the classes of levers, and explain how they make muscles more efficient.
3. Predict the actions of a muscle on the basis of the relative positions of its origin and insertion, and explain how muscles interact to produce or oppose movements.
4. Explain how the name of a muscle can help identify its location, appearance, or function.
5. Identify the principal axial muscles of the body, plus their origins, insertions, actions, and innervation.
6. Identify the principal appendicular muscles of the body, plus their origins, insertions, actions, and innervation, and compare the major functional differences between the upper and lower limbs.
7. Identify age-related changes of the muscular system.
8. Explain the functional relationship between the muscular system and other body systems, and explain the role of exercise in producing various responses in other body systems.

Teaching Strategies

1. Lecture and Lab Ideas

a. Unschooled in Latin and Greek, as are most of us, most students find muscle names to be virtually meaningless and thus, very difficult to learn. Emphasizing that there is logic to the nomenclature of muscles can help manage some of that difficulty. Project Table 11–1. Say the terms out loud with the students, so that they practice what the words sound like. Select terms whose equivalent meaning in English is likely to be known. Coax them along. Do they see the origin of our word *chin* in the Latin *genio-*? Dissect the words to point out familiar word elements. Point out that names based on relative location often occur in pairs. Thus, if there is an *externus*, there likely will be an *internus*; if there is a *superior*, one can expect an *inferior*.

b. Point out that the eye movements produced by the *superior* and *inferior obliques* are opposite to what the names may imply due to their mode of insertion. Point out also the pulley-like sling (*trochlea*) around which the superior oblique wraps to change the direction of its action. The *trochlear* nerve (N IV) innervates it.

c. Demonstrate the principle that muscles must cross a joint in order to cause the movement of a bone. As the appendicular musculature is examined, point out that the body (belly) of the muscle generally is found proximal to the bone that is moved and that the tendon of insertion spans a joint. For example, the muscles that move the upper arm are on the shoulder and chest; the muscles that move the forearm are on the brachium; and those that move the palm and wrist are on the antebrachium.

d. Point out that flexors of the elbow tend to be on the ventral surface of the arm, whereas extensors tend to be on the dorsal surface of the arm. Then point out that because our knees bend in a different direction than our elbows, in the legs, the flexors tend to be on the dorsal surface, whereas extensors tend to be on the ventral surface. In addition, abductors are generally on the lateral surface, and adductors on the medial surface.

e. Most muscle actions operate through third-class levers, by which a short muscle contraction is turned into a larger movement of the load. The biceps brachii flexing the forearm lifting a load held in the hand is a familiar example. Why third class? This leverage system overcomes a problem with muscles as machines (their limited shortening ability) but exploits a positive attribute (myofibrils can develop large forces).

f. A subtle design constraint on muscles that leads to the unipennate architecture is the fact that the risk of breakage in a long object (a thread, a rope, a muscle fiber) is a linear function of length. Because muscle fibers can't heal if broken, it pays to keep fiber lengths on the short side by building muscles with extended tendons of origin and insertion on opposite sides of the muscle belly, arranged such that the strong tendon becomes an appreciable fraction of the muscle length. *Semitendinosus* and *extensor digitorum longus* are good examples of this design principle.

g. Synergists can kick start a movement as well as stabilize a proximal joint. *Teres major* initiates arm movement continued by *latissimus dorsi*.

2. Analogies

a. Antagonistic muscle pairs are like the spring-loaded hinges on opposite sides of swinging doors.

b. Antagonistic muscle pairs are, in a sense, like the pairs of hydraulic actuators that move the bucket of a power shovel. These actuators operate the opposite of muscle; they only push!

3. Demonstrations

a. When discussing the sternocleidomastoid, instruct the students to turn their head to look sideways to make this muscle more visible. Have them palpate the two places of origin, the two muscle bellies, and the insertion. Encourage them to relate the name (sterno + cleido → mastoid) with the anatomy. This muscle is best demonstrated with a lean, well-muscled person. With the student's permission, the instructor should first perform the demonstration with a suitable volunteer. This approach can be used to demonstrate many other superficial muscles.

b. Point out, because of its insertion on the radial tuberosity, that biceps brachii both flexes *and* supinates the forearm.

c. An easy-to-understand demonstration of antagonistic muscle pairs is to show how gastrocnemius and tibialis anterior work to extend (plantar flex) and flex (dorsiflex) the ankle.

d. The calcaneal tendon through which the gastrocnemius and soleus muscles insert onto the calcaneus is made easily visible if the foot is dorsiflexed. (This manipulation will be useful later in demonstrating the calcaneal tendon reflex.) This tendon is commonly called the Achilles tendon after the character in Greek mythology whose mother held him by his heel as she dipped him into the River Styx to make him invincible. As luck would have it, he took an arrow in the heel, and perished.

4. Applications

a. The rectus abdominis muscles are referred to informally as the "abs" or the "six pack."

b. The pectoralis major muscles are referred to informally as the "pecs."

c. The latissimus dorsi muscles are referred to informally as the "lats."

d. The trapezius muscles are referred to informally as the "traps."

e. The scalenes (from vertebrae to ribs) are accessory muscles of inspiration [Figure 11–10 (a), (b)]. These are the muscles that stand out in the neck of a patient in respiratory distress struggling to inhale.

5. Common Student Misconceptions/Problems

a. Although learning the names and locations of the skeletal muscles is a difficult task, the danger is that mastering this learning task may divert resources from understanding how particular muscles and muscle groups produce specific movements. Because knowing the origin and insertions makes predicting the actions almost automatic, learning the tendinous attachments is essential. The textbook incorporates detailed resources (especially tables such as Tables 11–2 and 11–3, etc.). Another useful tool is an articulated skeleton (e.g., cat) when used as a companion to the preserved cat used in dissection studies. Also somewhat useful are the "dissectible" plastic models of the upper and lower extremities or an entire cadaver. Finally, palpation of the ropelike tendons in their own bodies can be helpful.

b. Students may be skeptical when you tell them that muscles only pull and that muscles never push. Learning the sliding filament mechanism for sarcomere shortening may not cause them to surrender their commonsense belief that muscles can push; after all, they push open doors and push the buttons on the TV remote. This is why the discussion of levers is essential. Upon learning the actions of levers, it becomes obvious how levers can change the direction of an applied force, turning a pull into a push.

6. Vocabulary Aids

a. Associate the word *rectus* with the word *erect*, that is, standing *straight* up.

b. The *corrugator supercilii* corrugates the brow (recall corrugated cardboard). This muscle's action produces a *supercilious* look (showing haughty disdain).

c. Note that the medial and lateral pterygoideus muscles are named for their origin on the pterygoid ("winged") processes of the sphenoid bone. The *ptero*dactyl was a winged lizard.

d. When discussing the muscles of the tongue, note that the Greek root *gloss* means "tongue." The cranial nerve that innervates these is the hypo*gloss*al (literally, "under the tongue"). *Glossolalia* is "speaking in tongues."

e. The *levator scapulae* "levitates" (lifts) the scapula (like shrugging your shoulders).

f. The *serratus anterior* has a sawtooth edge, like a "serrated" knife. *Serra* is Latin for saw.

g. The word *deltoid* describes the shape of this muscle, coming from the Greek capital *delta* (Δ), triangular in shape.

h. Stress that the *supraspinatus* and *infraspinatus* muscles are named for their location above and below the spine of the *scapula*, not because they are located on the backbone.

i. When discussing the *teres major* and *teres minor*, point out that the word *teres* means something long and round.

j. The *tensor fasciae latae* inserts on and tenses the iliotibial tract, stabilizing the leg laterally; it appears to have a long sheet of tendon (fascia). One can be reminded of the muscle's name by "The tensor fascia latae has a lotta fascia."

k. Point out that the *obturator* muscles have their origins on the obturator foramen.

l. The biceps femoris, the semimembranosus, and the semitendinosus compose the hamstring group; these three muscles are portions of the cut of meat sold as ham (which is the upper hind limb). Figure 11–20(c) makes this clear.

m. The sartorius is the longest muscle in the body. It is named for the Latin word for tailor, *sartor*. When sitting on the floor with legs crossed as was common, the work the tailor was sewing rested upon sartorius, the tailor's muscle.

n. The gastrocnemius muscle gets its name because of its belly-like shape (*gastro-*, belly). –*nemius* derives from the Greek word for leg.

CHAPTER 12

Neural Tissue

Introduction

Unit 3, "Control and Regulation," begins with this chapter about the fundamental properties of nerve cells, and ends with the endocrine system. Understanding nerve cells is central to understanding what it means to be alive. Although the topic might seem dull to the uninitiated, in fact, the nervous and endocrine systems underlie much of what students are concerned about, and this should be made clear to them. Learning, memory, the actions of therapeutic and "recreational" psychoactive substances, sensory experience, movement, neurological and psychological illness, growth, metabolism, and many other issues are linked to the basic science of the next six chapters. Chapter 12 introduces the basic properties of neural tissue, neurons, and neuroglia. Now is the time to challenge your students to explore how ionic currents, transmembrane potentials, nerve impulses, neurotransmitters, synapses, uptake inhibitors, and so on, terms they may have heard only in passing, ultimately add up to consciousness. To deepen your appreciation and understanding of the topics in this chapter, consult Eric Kandel's *Principles of Neural Science* (McGraw-Hill, 2000). More advanced treatments of excitability can be found in monographs such as Bertil Hille's definitive "Ion Channels of Excitable Membranes," 3rd ed. (Sinauer Associates, 2000).

Instructional Goals/Learning Outcomes

1. Describe the anatomical and functional divisions of the nervous system.
2. Sketch and label the structure of a typical neuron, describe the functions of each component, and classify neurons on the basis of their structure and function.
3. Describe the locations and functions of the various types of neuroglia.
4. Explain how the resting potential is created and maintained.
5. Describe the events involved in the generation and propagation of an action potential.
6. Discuss the factors that affect the speed with which action potentials are propagated.
7. Describe the structure of a synapse, and explain the mechanism involved in synaptic activity.
8. Describe the major types of neurotransmitters and neuromodulators, and discuss their effects on postsynaptic membranes.
9. Discuss the interactions that enable information processing to occur in neural tissue.

Teaching Strategies

1. Lecture and Lab Ideas

a. State clearly that the nervous system works together with the endocrine system to control and regulate all other systems of the body. The two systems, by virtue of their cellular processes, work on very different timescales. The nervous system is quick to respond, and its effects are short-lived. In contrast, the endocrine system is slower to respond, but its effects are longer lasting. The marriage of the two systems, the *neuroendocrine* system, provides a continuum of control.

b. The anatomy of a neuron as depicted in Figure 12–1 is not terribly complex but it is jammed with dozens of unfamiliar terms. So, take it easy when introducing this information in class. Point out what is familiar and groan along with them about the unavoidable multiplication of terms for familiar cell parts. Point out that dendrites look like trees with branches. Also, clarify that a neuron is both pre- and postsynaptic because the nervous system relies on chains of neurons, that is, neural circuits.

c. In a unipolar (sensory) neuron, the dendrites end and the axon begins without an interposed soma. This is shown in Figure 12–3. Sensory information moves in afferents as action potentials and not as local graded potentials; thus, it makes no physiological sense to label the portion of the afferent fibers distal to the dorsal root a "dendrite," as some texts do. Also of interest: The myelination changes from Schwann cells to oligodendrocytes in the region of the ganglion cell within the dorsal root.

d. Underscore the connection between the gross color of CNS tissue to its underlying structure and function: gray matter ⇒ high concentration of neuron somas ⇒ region of integration, while white matter ⇒ high concentration of myelinated axons ⇒ region of nerve conduction.

e. Make it clear that the channels in the presynaptic membrane that lead to release of acetylcholine are voltage-gated calcium channels, while the channels in the postsynaptic membrane that are opened by acetylcholine are chemically gated sodium–potassium channels.

f. One can never speak too often about integral membrane proteins. They underlie much of how cells and tissues work, and yet they remain a vague and distant notion to many students. In this chapter, voltage-gated channels, chemically gated channels, mechanically gated channels, and diverse pumps and transporters all crop up repeatedly in discussions of neural tissue. Link these to the larger topic of integral membrane proteins.

g. Get the students to equate the left- and right-hand sides of this table:

INPUT	OUTPUT
afferent	efferent
sensory	motor
dorsal	ventral (cord)
dendrites/soma	axon (at the cellular level)

If you put up just one side of a given row, can they supply the missing term?

h. Note that ependymal glia are epithelial cells within the CNS; they line the ventricles of the brain and central canal of the spinal cord. They possess many epithelial properties: continuous sheet of tightly packed cells that secrete and possess cilia.

i. Neuroglia greatly outnumber neurons but the neurons undertake the central roles of neural tissue, namely the storage and processing of information.

2. Analogies

a. It can be helpful to compare the nervous system to a telephone system. The CNS is the switchboard, the wires are the axons, and the telephones are the periphery. Point out, however, that unlike wires, axons carry information in ONE direction only. Since most students have cell phones, this analogy is almost obsolete!

b. A more modern and apt analogy to the nervous system, of course, is the digital computer. The advantage here is that information processing and data storage are what both do. The CPU is the CNS. Its transistors are its neurons, and they DO fire in a regenerative, all-or-none fashion. There is both short-term memory (DRAM, hippocampus) and long-term memory (hard drive, cerebral cortex). Inputs (afferents) such as the keyboard or mouse are clearly separate from outputs (efferents) such as the printer or monitor.

c. One can draw an analogy between the separate paths of information transfer to the movement of trains. To avoid collisions and keep things well managed, trains enter (afferents) the *rear* of the station (the dorsal root). After suitable "processing," they exit (efferents) the *front* of the station (the ventral root).

d. Chemically gated channels that underlie synaptic transmission can be likened to a key and lock. The neurotransmitter is the key; it has a very specific structure that permits entering the keyhole (sticking to the binding site) and opening the lock (gating the channel open). An agonist is like a master key; it can enter and turn the lock because it sufficiently resembles the structure of the key. An antagonist is like a faulty key that can enter a lock but not turn it. While it is in the keyhole, no other key can enter and so the channel is blocked.

e. Temporal summation of synaptic potentials is like bailing out a slowly leaking boat. If you bail slowly, you never can empty the boat; if you bail quickly, you can finally get the boat dry. This complements the textbook's analogy of summation to filling a leaky bathtub.

f. Many analogies can be used to illustrate the all-or-none principle beyond the example given in the text of the firing of a gun where the energy released by the bullet is independent of the pressure on the trigger. The flushing of a toilet is particularly apt: Nothing happens as the handle is pressed until a threshold is reached. Thereafter, the amount of water that is released is independent of how hard the handle was pressed (all-or-none behavior). The toilet cannot be flushed again until the tank refills (refractory period).

g. The propagation of the action potential is like the flame sizzling along a line of gunpowder. The wave of flame moves as the heat of the flame itself ignites the unburned powder. Thus, a local reaction causes the wave of activity to propagate along the axon (local inward current depolarizes the adjacent axon segment triggering its sodium channels to open).

3. Demonstrations

a. The "wave" performed at football games may be used to illustrate the continuous propagation of the action potential. The stimulus must be strong enough to get a minimum number of participants involved (exceed threshold). It propagates at a constant rate based on the delays of seeing and standing. It shows a refractory period in that a fan can't stand up again until he has sat down. It emulates continuous propagation in a related way: The "wave" moves but the people stay in place. The stimulus to stand is the standing up of spectators just to the side of you. You can try this in class—up one row and down the other as students rise when triggered by their immediate neighbor rising and sit when they have finished standing.

b. Sit on the edge of a table, and tap your patellar tendon with a reflex hammer or the like. The brisk knee jerk reflex contraction gives a dramatic introduction to much of what the section on the nervous system is concerned with. It demonstrates the *rapidity* of nervous system function.

c. Roll out a thin sheet of yellow modeling clay (the myelin-forming region of a Schwann cell). Form a cylinder the same length out of blue clay (the axoplasm). Wrap the yellow sheet around the blue cylinder. Slice the "cake roll" transversely to demonstrate a myelinated nerve fiber. The myelin should make up 40% of the diameter for accuracy.

4. Applications

a. Many drugs act at synapses. Some are agonists (e.g., nicotine, muscarine); some are uptake inhibitors (e.g., Prozac® and cocaine); some are neuromodulators (e.g., caffeine); some act presynaptically (e.g., tetanus toxin).

b. Local anesthetics block the perception of pain by blocking the sodium channel and thus preventing the passage of sodium ions. They act on any peripheral axon, both afferent and efferent; thus, both sensation and movement are blocked. Remind your students that dental anesthesia makes them numb but *also* makes it hard to speak and impossible to whistle!

c. Peripheral nerve regeneration after a nerve has been cut can occur because axons grow back within the tunnels of Schwann cells that remain after Wallerian degeneration. Central axons don't regrow because their myelin layer is formed by oligodendrocytes that become disrupted after lesioning, leaving no trail to follow.

d. All the biogenic amine neurotransmitters work through G protein-coupled receptor mechanisms. Drugs that target one of those receptors may have wide effects on a range of synapses.

5. Common Student Misconceptions/Problems

a. A danger in showing a classification of neuron types, such as Figure 12–3, is that students might be misled to think they all are of equal significance. Emphasize that most neurons are multipolar. The others are rare and/or highly specialized.

b. It may help your students to form a working model of the nervous system by drawing a highly simplified schematic on the board. Start by drawing a box representing a receptor; from it, draw a unipolar, afferent, sensory neuron leading from the PNS to the CNS through the dorsal surface of the cord. Use a simple sketch to represent a cross section of the cord. Stress that the cell body of the sensory neuron is outside the CNS in a ganglion. Next, draw an interneuron. Continue by drawing the multipolar, efferent, motor neuron leaving the CNS within the ventral root to a second box representing the effector. Indicate that the effector could be a skeletal muscle, a smooth muscle, or a gland. Stress that the cell body of the motor neuron is within the CNS (in the gray matter of the cord). Reiterate the function of each neuron; emphasize the *unidirectional* flow of information. Point out the *synapses*. Try erasing parts of the pathway one by one, and ask the students, "If this were injured, what function would be lost?"

c. Students have difficulty understanding the origin of the electrical charge across the plasma membrane. They can see the ionic concentration differences but fail to see how that leads to a voltage difference. Emphasize that the numbers of cations leaving and entering are not equal. Potassium ions exit faster than sodium ions can enter because the resting membrane is much more permeable to potassium ions. Thus, the resting membrane potential is established more by the potassium efflux than by the sodium influx. With depolarization, sodium permeability rises markedly, leading to the net entry of positive charge. This excess positive charge reverses the membrane potential.

d. All peripheral axons, both unmyelinated and myelinated, are surrounded by Schwann cells.

e. Make it clear what information in Table 12–6 they are expected to know, if any.

6. Vocabulary Aids

a. When discussing astrocytes note that the word *astro-* means "star" and denotes the shape of these cells.

b. "Dendro-" is the combining form from Greek for *tree*. Students encounter this root in *dendr*ites, axo*dendr*itic, telo*dendr*ia, and oligo*dendro*glia.

c. To remember the sign of the resting membrane potential, associate ***N***egative with i***N***side, and p***O***sitive with the ***O***utside.

d. To distinguish afferent from efferent, remember ***E***xit means to leave. That will remind you of ***E***fferent. Afferent is the other one!

e. *Axons* don't think; they ***act!*** That is, axons carry *act*ion potentials.

CHAPTER 13

The Spinal Cord, Spinal Nerves, and Spinal Reflexes

Introduction

Considering its crucial role, the spinal cord is surprisingly small in cross section; in fact, the human gastrointestinal tract is said to contain more neurons than the spinal cord! Students are often curious about two important clinical questions that concern the cord, *regional anesthesia* (spinal blocks, epidural blocks) and *spinal cord injuries.* Both topics are familiar to students; some may have had the benefit of an *epidural block* and all are familiar with the grave consequences of spinal cord injury, in particular because of the paralyzing injuries suffered by the late Christopher Reeve, an actor and spokesperson for spinal cord injury (SCI) research. You might ask for a class volunteer to discuss their experience with epidural anesthesia, or as caregivers of SCI patients. This can then motivate a discussion of why, for the present at least, spinal cord injury is largely irreversible. This links this chapter to the previous chapter on neural tissue. Clinical material concerned with other practical issues such as meningitis, spinal taps, and shingles are highlighted in the chapter.

Instructional Goals/Learning Outcomes

1. Describe the basic structural and organizational characteristics of the nervous system.
2. Discuss the structure and functions of the spinal cord, and describe the three meningeal layers that surround the central nervous system.
3. Explain the roles of white matter and gray matter in processing and relaying sensory information and motor commands.
4. Describe the major components of a spinal nerve, and relate the distribution pattern of spinal nerves to the regions they innervate.
5. Discuss the significance of neuronal pools, and describe the major patterns of interaction among neurons within and among these pools.
6. Describe the steps in a neural reflex, and classify the types of reflexes.
7. Distinguish among the types of motor responses produced by various reflexes, and explain how reflexes interact to produce complex behaviors.
8. Explain how higher centers control and modify reflex responses.

Teaching Strategies

1. Lecture and Lab Ideas

a. It is essential that students "reflexly" distinguish dorsal (posterior) from ventral (anterior) and learn to associate sensory afferents with posterior roots and motor efferents with anterior roots. Have the students practice this identification with an unlabeled microscopic cross section. If the section includes the dorsal root ganglion, point out its cellularity. If elements of the meninges are present, exploit this in your lab discussion.

b. A simplifying principle for spinal pathways is this: *Ascending* tracts are always *sensory*; *descending* tracts are always *motor*. Point out that somatosensory information enters the dorsal aspect from the periphery and moves up the cord to the brain; except for segmental reflexes, motor responses move down from the brain and out the ventral aspect to the periphery. These topics are covered in detail in Chapters 15 and 16.

c. Spend a few minutes with Figure 13-1 providing a "roadmap" for the detailed study of the nervous system that lies ahead in this unit.

d. Using Figure 13–2 point out the 31 spinal segments (C, T, L, S, Cx) and the cranial–caudal gradient in the proportion of white matter to gray matter because of the increasing number of sensory and motor axons closer to the brain.

2. Analogies

a. Impulse conduction in the nervous system only works correctly if it's one way. This is like the proper operation of freeways. Traffic flow occurs in inbound or outbound lanes, separated by a barrier. Traffic going the wrong way in a one-way road is chaos! To change directions, information must negotiate a "cloverleaf exchange," that is, enter the CNS for processing by synaptic circuits.

b. Presynaptic facilitation or inhibition is like a conductor raising or lowering one hand while keeping the rhythm going with the other. The basic pattern doesn't change but the loudness does.

3. Demonstrations

a. Act out in slow motion the *crossed extensor reflex* as if you stepped on a sharp object (a Lego is a familiar household hazard). Ask the class what happens to the load on the other foot when one foot is raised. [The load on the other foot doubles!] Ask them what properties the reflex needs to handle that mechanical fact. Guide the students to the concluson that the reflex must be "wired" to increase excitation to the contralateral leg extensor motor neuron pools (recruit more motor units; generate more tension in the active units by discharging more rapidly).

b. Demonstrate the ability of descending influences to increase the excitability of quadriceps motor neurons (bringing them closer to threshold), by investigating the strength and briskness of the patellar stretch reflex (knee jerk) by repeatedly tapping the knee with a reflex hammer as the subject voluntarily contracts forearm muscles by making a tight fist.

c. Use a penlight to demonstrate the papillary reflex as an example of a visceral reflex. Show that it's bilateral in a healthy brain. Point out that visceral simply means nonskeletal muscle as the effector. It applies outside the viscera (the "guts").

4. Applications

a. Peripheral neuropathies can occur in the arms and hands of someone using ill-fitting crutches that put constant pressure on nerves within the axilla. The functional loss can become irreversible if it is not treated quickly.

b. Some students are aware that multiple sclerosis is an autoimmune process that disrupts and destroys the myelin on *central* axons, causing loss of sensation and movement. However, they are usually unaware that an exactly parallel disease can selectively damage *peripheral* myelinated nerves. One form, *Guillain–Barré* syndrome, has a sudden onset and can strike anyone irrespective of age or gender. The cause is unknown but often follows a viral or bacterial infection. This brings out the difference between myelination of CNS and PNS nerve fibers.

c. Erb's palsy, injury to the brachial plexus leading to weakness and loss of feeling in the shoulder and arm, frequently results from birth trauma. Excessive stretch of the shoulder can cause irreversible damage.

5. Common Student Misconceptions/Problems

a. Students often have difficulty understanding that the PNS and CNS are physically continuous. It may help to trace a sensory impulse from a cutaneous receptor along a peripheral nerve to the dorsal root, dorsal root ganglion, and then to the dorsal horn using Figure 13–7(b).

b. Students often equate afferent with ascending, and motor with descending. Make clear that *afferent* and *efferent* are terms that only apply to axons carrying information in the PNS. Reserve ascending and descending for sensory and motor pathways within the CNS.

c. Use Figure 13–7 (a) and (b) to systematically review the cross-sectional anatomy of the cord. Point out these essential features:

   1. Sensory axons enter by way of the dorsal root. Motor axons leave by way of the ventral root.
   2. Afferent cell bodies reside in the dorsal root ganglion. Efferent cell bodies reside in the lateral and anterior gray horn.
   3. Somatic afferents synapse in the dorsal horn but visceral afferents synapse in the intermediate horn.
   4. Somatic efferents have but a single peripheral synapse, whereas visceral efferents have two [show with Figure 13–7(a)]. Indicate that the white ramus carries the preganglionic myelinated axons and that the gray ramus carries the postganglionic unmyelinated fibers. Underline that this automatically implies that the ganglion lies close to the CNS in the sympathetic nervous system.

d. Students have a hard time dissociating action from intent. Let them know that spinal reflexes are based within the spinal cord and require no conscious awareness. We only become consciously aware of them from ascending tracts carrying somatosensory information to the cortex where consciousness is housed.

e. Students need to know concretely the level of detail they are expected to learn.

6. Vocabulary Aids

a. The literal translation of dura mater is "hard mother." It takes its name from the tough biomembrane that forms over fermenting wine, referred to as the "mother" of the culture.

b. The network of delicate fibers within the arachnoid membrane gives this layer a weblike appearance; *arachnoid* means "cobweb-like" in Greek.

c. The translation of pia mater is "tender mother," because it is so delicately thin.

d. Point out that the prefixes used to describe the connective tissue layers of nerves (epi-, peri-, endo-) are the same as those used to describe the connective tissue wrappings of muscles.

CHAPTER 14

The Brain and Cranial Nerves

Introduction

Mastering the detail in this chapter is beyond the scope of most undergraduate human anatomy and physiology courses. This implies that teaching this topic appropriately means identifying the key structural and functional information you want your students to carry away and to let them know explicitly what you expect. To identify the main points, review the chapter's Study Outline found near the end of the chapter. By sticking to the main points about brain anatomy and physiology, the danger that the forest will be lost among all the trees can be minimized. The anatomy in this chapter provides the basis for discussing the intricacies of brain function. So, excessive detail may confuse and dismay students, and thus compromise later study. Students may find it useful if you adopt the same approach as the text; begin with an overview of brain anatomy and follow that with a bottom-up presentation from the medulla oblongata up to the neocortex, that is, from smallest and simplest to largest and most complex.

Instructional Goals/Learning Outcomes

1. Name the major brain regions, vesicles, and ventricles, and describe the locations and functions of each.
2. Explain how the brain is protected and supported, and discuss the formation, circulation, and function of cerebrospinal fluid.
3. Describe the anatomical differences between the medulla oblongata and the spinal cord, and identify the medulla oblongata's main components and functions.
4. List the main components of the pons, and specify the functions of each.
5. List the main components of the cerebellum, and specify the functions of each.
6. List the main components of the mesencephalon, and specify the functions of each.
7. List the main components of the diencephalon, and specify the functions of each.
8. Identify the main components of the limbic system, and specify the locations and functions of each.
9. Identify the major anatomical subdivisions and functions of the cerebrum, and discuss the origin and significance of the major types of brain waves seen in an electroencephalogram.
10. Describe representative examples of cranial reflexes that produce somatic responses or visceral responses to specific stimuli.

Teaching Strategies

1. Lecture and Lab Ideas

a. The gyri and sulci are a bit random but a general pattern exists sufficient to name and identify particular features (e.g., central sulcus, dentate gyrus). On the other hand, the patterns can differ for the two hemispheres of the same individual. Point this out to your class with the real anatomical data of Figure 14–18, a view of the base of the brain.

b. Ask the students to trace a drop of CSF as it flows from the site of its production (choroid plexuses within the lateral ventricles) to the place of its reabsorption (arachnoid granulations in the superior sagittal sinus). Rely on Figure 14–4 to document this trip.

c. Use Figure 14–14(b) to illustrate the pattern of gray and white matter in the cerebrum. At this point, the names of specific nuclei and tracts are unimportant. Point out that the pattern in the spinal cord is gray on the inside and white on the outside, the reverse of what is seen in the cerebrum. Point also to the deep forebrain nuclei. Remind the students that wherever they see gray matter, they are seeing a region rich in neurons, and thus, a region for the interpretation and integration of information. The regions of white matter are myelinated tracts of axons leading to and from the centers.

d. It is the control of the cerebral nuclei that coordinates the unconscious swinging of our arms with our legs as we walk. Instead of swinging arms with opposite legs, as we unconsciously do when we walk, suggest to the students that they try swinging their right arm with their right leg and their left arm with their left leg. This apparently simple task turns out to be surprisingly difficult. This experience should let them better appreciate the "unconscious control" that the basal nuclei exert.

e. If you are standing in a large area with lots of people around, and someone calls your name, it is the reflex controlled within the corpora quadrigemina that enables you to turn your head in the direction of the sound to seek its origin. That is, visually orienting to the source of a sound is an ability housed in adjacent parts of the midbrain.

f. Point out the amazing fact that, of the twelve cranial nerves, six (II, III, IV, V, VI, VII) are involved with eye function.

g. Describe the two intrinsic muscles of the iris. Underscore that the oculomotor nerve (CN III) controls only the circular (sphincter) muscle of the iris (constriction) and not the radial muscle (dilation). Tell the students they should anticipate continued discussion of these smooth (visceral) muscles when covering the autonomic nervous system.

h. In addition to the mnemonic device given in the text for remembering the names of the cranial nerves in order, there is one that is frequently used to remember the primary functions of the cranial nerves: "Some Say Marry Money, But My Brother Says Bad Business Marry Money." (S = sensory, M = motor, B = both).

i. Return to Figure 14–1 before leaving the chapter. It is loaded with the key information of this topic. It clearly defines the major brain regions and their functions. Help them see the shadow of the diencephalon and brain stem in the overall figure.

j. Follow presentation of Figure 14–11 with Table 14–6, which verbally summarizes the litany of hypothalamic functions. Let them appreciate how much important stuff is jammed in a rather tiny volume of nervous tissue.

2. Analogies

a. The hierarchy of brain divisions is somewhat like the hierarchy of government. The smallest subdivision, a town or village, is the most crucial, affecting daily life at all times. As you move up the neural axis, it is like moving up to county, state, and national government. The higher up you go, the larger, more complex, and more redundant are the components.

b. Compare the two thalami to two scoops of ice cream side by side in a cone. The hollow cone is the third ventricle leading to the aqueduct of the midbrain.

c. Stress that the brain derives from the original embryonic neural tube and thus is hollow. The ventricles are filled with cerebrospinal fluid produced by specialized cells within ependyma of the choroid plexus. The ventricles are continuous with the central canal of the cord. To clarify these relations, take them on a ride on a miniature "Yellow Submarine" beginning in the most caudal corner of the lateral ventricle, noting the fingers of choroid plexus waving like seaweed, then down the interventricular foramen to the third ventricle, through the midbrain aqueduct to the shallow sea of the fourth ventricle to the central canal to the conus medullaris. Rely on Figure 14–2 to document the trip.

d. As shown in Figure 14–1, suggest that the brain is like a mushroom. The cap is the cerebrum; when it's pulled off, it reveals the diencephalon and brain stem beneath it.

3. Demonstrations

a. Show the corpus callosum in both sagittal and coronal views [e.g., Figures 14–10 and 14–14(c)] so that students can better visualize its size and extent. In lab, while dissecting the sheep brain, insert your thumbs into the longitudinal fissure and gently spread the two cerebral hemispheres apart to reveal its extent *in situ*.

b. Deep cerebral structures are more difficult for students to visualize, particularly in a three-dimensional sense. Use Figure 14–14 to help describe the location of the basal nuclei. Help them relate the sectional views in (b) and (c) to the reconstruction in (a). Indicate that the third ventricle and thalami are posterior to the plane of section in Figure 14–14(b). Point out that the head of the caudate nucleus is superior to and more medial than the lentiform nucleus, but the tail is inferior to the lentiform and lateral to the head. Thus, the caudate nucleus curls around the lentiform nucleus like a lock of hair curling around someone's ear.

c. As with the basal nuclei, the location of the limbic system may be difficult for students to visualize. State that the limbic system is a functional rather than an anatomical grouping. Before showing Figure 14–11, begin with the list in Table 14–7. Having a concrete list of parts and functions beforehand helps in interpreting the complex anatomy. This is a good place to use double-projection systems where available. The limbic system, viewed bilaterally, resembles the horns of a bighorn sheep. Figure 14–11(b) provides a three-dimensional reconstruction.

d. To describe the left and right thalami, ball your hands and place them together, knuckles touching, to indicate how close the walls of the third ventricle are to each other. Describe the thalamic nuclei as being located within each hand.

e. One method to remember the five facial regions served by the facial nerve is to place the palm of the hand over the ear, and then spread the fingers to touch each of the five regions: temporal, zygomatic, buccal, mandibular, and cervical.

f. To dramatize the limit of our understanding of higher brain function, point out how much neural real estate of the cerebral cortex is labeled "association" or "integrative," about which we know almost nothing in detail.

4. Applications

a. The introduction of ionized compounds into the CNS (e.g., for chemotherapy of cancer or infections) cannot rely on intravenous injection because of the blood–brain barrier. In this case, drugs are injected "intrathecally" (literally, "within a sheath") by lumbar puncture into the subarachnoid space. The rapid circulation of the CSF distributes the medication widely within the brain.

b. Branches (maxillary, mandibular) of the trigeminal nerve (CN V) often are the nerves that are "deadened" with local anesthetic by the dentist.

c. The dural folds can act like knives to slice into brain substance. To-and-fro violent movement (e.g., shaken baby) can cause the falx cerebri to slice into the brain. Likewise, side-to-side movements can cause damage by the tentorium cerebelli.

d. Neural tube defects (e.g., spina bifida, hydrocephalus) result from improper formation of the early embryonic nervous system. Including folic acid in the diet prevents many cases. Since the neural tube forms before the first missed period, any woman of child-bearing potential should consume adequate levels of folic acid (400 μg/day). Ask the students what foods are good sources of folic acid. Accumulate their responses on the board (enriched grain products, flour, orange juice, dark green leafy vegetables, beans, nuts, poultry, liver).

e. When reviewing Figure 14–10, point at the cut infundibulum. The pituitary is always missing because it tears off in the sella turcica when the brain is removed at autopsy. In years past, the NIH would collect human pituitaries to prepare growth hormone to treat pituitary dwarfs.

5. Common Student Misconceptions/Problems

a. Due to its complexity, brain structure and function present a substantial challenge for most students. Part of the difficulty is to visualize that the brain is a three-dimensional structure with cavities inside, and not a solid homogeneous mass. To help with this, allot some time to the embryological development of the brain so that students realize that the brain originates as a fluid-filled neural tube that transforms to the three primitive brain vesicles. Emphasize that the "hollowness" continues throughout development, eventually becoming the four fluid-filled ventricles. Rely on Figure 14–2; it shows a three-dimensional reconstruction in both lateral and posterior views.

b. Students see some "logic" to the location of the auditory cortex in the temporal lobe because of the spatial relationship to the ears, but the location of the visual cortex seems odd. This should caution them not to place too much confidence in "common sense."

6. Vocabulary Aids

a. The names of the five secondary regions of the brain will make more sense if the meanings of the Greek combining forms are given. Mention first that the Greek root *encephalo-* refers to the brain (literally, it means "inside the head"). From telophase (the distant or end phase) in mitosis, they know *telo*. Thus, the *tel*encephalon is the top end of the CNS. The brain then "passes through" (*dia-*) the *di*encephalon to the brain stem. The brain stem is made of the "middle" (*mes*encephalon), the part "beyond" (*met*encephalon—recall that the *meta*carpals are beyond the carpals), and the part (*myel*encephalon) adjacent to the "marrow" of the vertebral column, that is, the spinal cord.

b. Associate the "S" in the word **S**ensory with the "S" in po**S**tcentral gyrus.

c. Point out that the name of the corpora quadrigemina means "four twins," namely, the four colliculi (hillocks).

d. Limbic comes from *limbus,* Latin for "border." This grouping is at the *border* of cerebrum and diencephalon.

e. Before describing the function of the *trochlear* nerve (CN IV), remind the students of the trochlea of the humerus, so called because it resembles a pulley. Given that reminder, ask them to figure out which extrinsic eye muscle the trochlear nerve controls.

f. As you discuss the abducens nerve (CN VI), ask the students what the word *abducens* reminds them of. Prompt them into saying "abduct." Ask them how the eyeball would move if it were being "abducted." Then ask them to deduce what extrinsic eye muscle this nerve controls.

g. Associate the word *vagus* (the wandering nerve) with the word *vagrant,* a wandering person. The vagus is the only cranial nerve that leaves the head region, and "wanders" to the viscera of the thorax and abdomen, far from the other cranial nerves.

h. The textbook has dropped the use of the term "basal ganglia." It reserves the term *ganglia* exclusively for PNS structures. The term is described as being an older, more inclusive term. Modern usage has moved in this direction but your students are likely to encounter "basal ganglia" in their professional work or further studies. Thus, you should mention it, and for ease of presentation, identify it with the basal nuclei, although more caudal gray matter has been included in past descriptions. The rare soul who needs to know this (such as this writer) will have to learn it on their own.

i. *Falx* means "scythe." Although named for its shape, it aptly reminds us of the potential of the falx cerebri for slicing into the brain during sudden front-to-back motions of the head.

CHAPTER 15

Neural Integration I: Sensory Pathways and the Somatic Nervous System

Introduction

Whereas Chapters 13 and 14 were largely concerned with spelling out the details of CNS anatomy on a regional basis with only limited attention to global functions, Chapter 15 integrates sensory and motor function along the entire neural axis. That is, this chapter provides a basis for understanding sensory experience and motor behavior. It is important to make students aware that we do not as yet have a deep understanding of these central elements of consciousness. This should not surprise us: many neuroscientists regard the human brain as the most complex object in the solar system. This topic provides a good opportunity to discuss the limitations of science. The integration of sensory information from many sensory receptors to form an apparently seamless representation of reality, along with an apparently effortless ability to plan and execute movement and to recall distant experiences, are almost entirely unexplained. Physician and scientist V. S. Ramachandran and writer Sandra Blakeslee offer a masterful account of observational and experimental approaches to studying these thorny problems in *Phantoms in the Brain: Probing the Mysteries of the Human Mind* (Harper Perennial, 1999). This book is suitable both for students and teachers. Some of his simple demonstrations could be used in a lab; it could also serve as a source of topics for special research reports.

Instructional Goals/Learning Outcomes

1. Specify the components of the afferent and efferent divisions of the nervous system, and explain what is meant by the somatic nervous system.
2. Explain why receptors respond to specific stimuli, and how the organization of a receptor affects its sensitivity.
3. Identify the receptors for the general senses, and describe how they function.
4. Identify the major sensory pathways, and explain how it is possible to distinguish among sensations that originate in different areas of the body.
5. Describe the components, processes, and functions of the somatic motor pathways, and the levels of information processing involved in motor control.

Teaching Strategies

1. Lecture and Lab Ideas

a. After explaining the "from–to" naming scheme for the sensory and motor pathways, interactively quiz the students by giving them names of pathways and asking them to decide: (1) if it is ascending or descending, (2) if it is sensory or motor, and (3) what origin and destination are indicated in the name.

b. Thermal receptors detect temperature both statically and phasically. However, different receptors respond best to either heating or cooling.

c. Use Figure 15–1 to provide a road map to this and the next chapter. The figure clarifies the cross talk between sensory and motor systems, and will help the students to see how input–output relate to higher function covered in Chapter 16.

d. Point out that because the dorsal root is the afferent pathway from the periphery, it makes sense that the dorsal columns should be the sensory tracts within the CNS. Similarly, point out that, with the exception of the rubrospinal and lateral corticospinal tracts, the descending motor pathways occur in the ventral region of the spinal cord. This places them close to the ventral roots, the efferent path to the periphery.

e. Note that the four cranial nerves that send afferent fibers into the medial lemniscus are the same four cranial nerves that are classified as "mixed."

f. Stress that proprioceptive information carried along the spinocerebellar pathway terminates in the cerebellum. Consequently, we have no conscious awareness of the sensations generated by muscle spindles and tendon organs because this pathway does not reach the cerebral cortex. Some joint receptors do project to the somatosensory cortex and so are perceived.

g. The midbrain reticular formation can focus our attention on particular sensory modalities; thus, the brain can select among the streams of sensory information, attending to some, ignoring others.

2. Analogies

a. Compare the inhibitory effect on excitatory motor neuron activity to a radio that is switched on but with the volume turned way down. The signal is there, but the sound is being inhibited. If the inhibition were reduced (the volume turned up), we would hear the radio louder.

b. The senses are like newspaper stories. They answer the questions, What (is the stimulus?); Where (is it located?); How (strong is it?).

c. The basal nuclei do not send axons to motor nuclei but only to the thalamus and brain stem reticular formation. These inputs affect movement by increasing or decreasing feedback loop gain. When loop gain is too high, tremor occurs. This is analogous to oscillation in an audio system when there is "too much feedback."

3. Demonstrations

a. With their eyes closed, in a quiet place, have a student lightly touch another with an ice cube. [Use a towel to catch drips.] Can the subject distinguish the different arrival times of the light touch sensation (carried over the fast-conducting dorsal column pathway) and the cold sensation (carried over the slower spinothalamic pathway)? Results will be improved if they are prepped with the knowledge of what they might experience. Use a simple fast and slow path diagram to show the arrival of different modalities along different sensory pathways but arriving at the same locale on the somatosensory cortex.

b. Have the students enumerate the number of synapses in the two main somatosensory pathways, diagram where they are found, and point out whether the pathway crosses the neural axis. This will demonstrate an underlying similarity.

c. Using drafting dividers [two sharp points on two radii], do simple two-point discrimination trials. Compare first phalanx to second. Ventrum of forearm to dorsum. This never fails to fascinate students and is readily explained in terms of innervation density and receptive field size. They might also appreciate that the skin is a sensory mosaic of sensitivities and modalities.

4. Applications

a. Amyotrophic lateral sclerosis (ALS), often referred to as *Lou Gehrig disease*, is named after the beloved New York Yankees first baseman who died of ALS in 1941. The disease remains incurable and fatal. Lower motor neurons die, leading to complete flaccid paralysis and respiratory failure.

b. If you are not afraid of dating yourself, you can ask the students if they are familiar with the music of Woody Guthrie, the American folk musician who died of Huntington disease. Point out that it has been possible to diagnose the disease by direct mutation analysis since 1993. Make your students aware of OMIM, the *Online Mendelian Inheritance in Man* database of genes and disease. OMIM is a rapidly expanding treasure trove of information applying genomics, proteomics, and genetics to a range of human traits.

c. Alzheimer disease can't be diagnosed reliably in its early stages; only postmortem examination of brain sections under the microscope that reveals the characteristic plaques and neurofibrillary tangles is definitive. The development of a test to detect an early stage of the disease offers the prospects of halting, or even reversing, its progression.

d. Fast-adapting cutaneous mechanoreceptors can be specifically stimulated in a neurological exam by the vibrations of a tuning fork (440 hertz is about a two-millisecond cycle length).

e. The sharp point (pain) and blunt side (light touch) of a safety pin can be used to probe the skin to localize and differentiate spinal cord injury, with the dermatomal map in mind.

f. The fact that morphine (synthesized by a plant) can block pain in the spinal cord and brain implies that there are CNS receptors of some kind for morphine and ligands, which are the endogenous, physiological agonists. These concepts fostered intensive worldwide research that led to the discovery of the endorphins and enkephalins, of the molecular identity of the endorphin receptor family, and to characterization of the essential structural features of endogenous and foreign agonists. Powerful new pain medicines have resulted from these insights.

5. Common Student Misconceptions/Problems

a. Students often get confused about tracts and pathways. Emphasize that more than one synaptic relay may be involved in a tract or pathway. Also, reiterate the "from–to" convention on naming that permits immediate distinction between ascending and descending tracts.

b. Students often equate afferent with ascending and efferent with descending. Make it clear that afferent and efferent apply only to the PNS.

6. Vocabulary Aids

a. When discussing nociceptors, note that the Latin word *noxa* (meaning, harm) is the root of "noxious." Nociceptors signal real or potential *harm* to the tissue they monitor.

b. When discussing the fasciculus *cune*atus, remind the students of the combining form "cuneo-" that they learned when studying the wedge-shaped metatarsal *cunei*form bones. It might remind them of *cunei*form script, wedge-shaped writing pressed into clay tablets in ancient Sumeria.

c. Likewise, the narrower fasciculus *gracilis* should remind them of the slender gracilis muscle on the medial side of the thigh. Gracile means slender or thin.

d. The "***P***" in substance P (which actually stands for peptide) can be mnemonic for ***P***ain, which it transmits ***P***eripherally.

e. *Somato*- refers to body (*soma*). *Somato*sensation is the body sensation, that is, sensation of the body, excluding the special senses.

f. Ironically, "general" senses means something quite specific: all the senses that are not special.

g. The term "receptive field" can apply in the case of the special senses as well as the tactile sense. For example, a region within the visual field is the receptive field of a given ganglion cell.

h. Note in Figure 15–3 how few eponymic names for the bestiary of cutaneous tactile receptors survive in modern usage.

i. The term *extrapyramidal system* is passé. Our present understanding of motor control is based on cooperation and interfunctionality of three main motor systems. They are summarized in Table 15–2. It would be useful to show this as a summary to your presentation on somatic motor control.

CHAPTER 16

Neural Integration II: The Autonomic Nervous System and Higher-Order Functions

Introduction

The authors begin their narrative with the observation that, "Your conscious thoughts, plans, and actions represent a tiny fraction of the activities of the nervous system." This focuses attention on the essential role played by the nervous system in the *unconscious* second-by-second maintenance of a constant internal milieu. That is, the responsibility for defending homeostasis falls largely on the autonomic nervous system. Therefore, it seems inevitable that many useful medicines target the autonomic nervous system. Students may be acquainted with some of these. You might start with a list of common diseases (hypertension, asthma, urinary leakage or insufficiency) and ask what sorts of medicines are used to treat them (often, adrenergic and cholinergic agonists and antagonists). At this point, few will know how these drugs work. Promise them that by the time this chapter is finished they should (and are expected to) be able to discuss the normal physiology of the autonomic nervous system and understand how familiar medicines work to modify ANS function.

Instructional Goals/Learning Outcomes

1. Compare the organization of the autonomic nervous system with that of the somatic nervous system.
2. Describe the structures and functions of the sympathetic division of the autonomic nervous system.
3. Describe the mechanisms of neurotransmitter release and their effects on target organs and tissues.
4. Describe the structures and functions of the parasympathetic division of the autonomic nervous system.

5. Describe the mechanisms of neurotransmitter release and their effects on target organs and tissues.
6. Discuss the functional significance of dual innervation and autonomic tone.
7. Describe the hierarchy of interacting levels of control in the autonomic nervous system, including the significance of visceral reflexes.
8. Explain how memories are created, stored, and recalled, and distinguish among the levels of consciousness and unconsciousness.
9. Describe some of the ways in which the interactions of neurotransmitters influence brain function.
10. Summarize the effects of aging on the nervous system.
11. Give examples of interactions between the nervous system and each of the other organ systems.

Teaching Strategies

1. Lecture and Lab Ideas

a. Considering all the fine-scale detail presented in the preceding several chapters, this is a good time to get our bearings again. Figure 16–1 lets you show them where they have been and where they are going.

b. Use Figure 16–2 to bring out the key differences between the somatic and visceral nervous systems: that the somatic system drives skeletal muscles directly whereas the visceral nervous system drives smooth muscle and glands indirectly through peripheral ganglia.

c. Emphasize that ANS effects on peripheral effectors may be either to activate *or* inhibit them, in contrast with the purely excitatory somatic nervous system. It is this property that enables control by dual innervation, for example, airway resistance, heart rate, and pupil diameter.

d. To make the "fight or flight" stimulus for sympathetic activation more familiar, call it the "blue light syndrome." This is the constellation of symptoms we experience when a police officer turns on his blue light and accelerates toward our car. Help them make a list (rapid pounding heartbeat, sweaty palms, deep breathing, dry mouth, dilated pupils). You might mention that in 1929 the American physiologist Walter Cannon characterized sympathetic activation as preparing for "fight or flight." This could be updated to "kick butt or haul ass," if this phrase won't offend the more sensitive members of your class.

e. Regarding the "craniosacral" organization of the parasympathetic nervous system point out that only four cranial nerves (III, VII, IX, and X) compose, the "cranio" portion of the PS branch of the ANS and the "sacral" portion involves only S_2, S_3, and S_4. The cranial nerves innervate ganglia in the head, neck, thorax, and abdomen (remind them of the "vagrant" nature of the vagus). "Rest and digest" (an update of "rest and repose") can be used to simply characterize parasympathetic function.

f. Figure 16–2 displays a side-by-side structural comparison of the sympathetic and parasympathetic branches of the ANS for which Table 16–2 makes a useful detailed companion. Functionally, the systems are compared in detail in Table 16–3. Table 16–3 combines in words what is diagrammed anatomically in Figure 16–5 (sympathetic) and Figure 16–8 (parasympathetic). These three graphic elements concisely summarize much of the chapter's information about the ANS.

g. Using Figure 16–4 trace the preganglionic and postganglionic fibers as they pass through the white and gray rami communicantes. Point out that the preganglionic fibers are white because they are myelinated; the postganglionic fibers are gray because they are unmyelinated.

h. The innervation of the suprarenal medulla by preganglionic fibers is a direct link between the nervous and endocrine systems. Most students know that adrenaline (epinephrine) is released during times of stress; they could probably describe the symptoms of an "adrenaline rush."

i. Ask the students to explain how a single neurotransmitter, for example norepinephrine, can be stimulatory to some targets but inhibitory to others. Answer: different adrenergic receptors and second messengers. The same principle applies to cholinergic transmission (nicotinic, muscarinic). See Table 16–1 for a summary.

2. Analogies

a. Each autonomic ganglion is like the handoff in a relay race. The pre-ganglionic fiber "hands off" the baton to the postganglionic fiber that carries the race to the target effector. In the sympathetic NS, the "baton" changes from acetylcholine to norepinephrine.

b. The stages of sleep are like living in the subbasements in a large building. On a regular basis, we climb down, spend a while, climb down again, spend a while, and then head back up.

c. The conversion of short-term memory is like turning quickly scribbled lecture notes into properly word-processed files, suitable for long-term storage and recall.

d. The constant random bombardment of the reticular formation with sensory signals from many modalities to keep the cortex "up" is like the many hands in a group keeping a beach ball *up* in the air.

3. Demonstrations

a. Adrenergic agonists are widely used to modify illnesses. List the warnings against use of the behind-the-counter medicine pseudoephedrine (e.g., hypertension, diabetes, hyperthyroidism). Ask the students why the use of the medication is contraindicated for people who have these conditions. Why are dry mouth or racing of the heart common side effects?

b. Remind them that long-term memory (the kind you want them to form in this course!) is largely permanent. To demonstrate this cardinal fact about learning, ask them to recall where they were and what they were doing on the morning of September 11, 2001. How about the morning of the day before? Why the difference in the recollection? [Powerful experience, often recalled.]

c. Ask, "Why do loud noises and bright lights rouse you from dozing?" Also ask, "Do you see why emergency services and fire alarms rely on these modalities?"

4. Applications

a. Impress upon those students planning on entering health care professions they should try to understand the central role the autonomic nervous system plays in maintaining homeostasis. Because most diseases either lead to or result from disruptions to autonomic function, this understanding of unconscious control is crucial for comprehending normal and pathological physiology.

b. The consensual pupillary reflex is routinely used as a quick evaluative tool during a neurological exam, especially if injury to the midbrain is suspected. This reflex is easy to demonstrate with a penlight in a partly darkened classroom or lab. This always wows the class. Simple but powerful. If absent, midline tectal structures have been injured!

c. *Belladonna* (literally, "beautiful lady") is a toxic plant once used for *cosmetic* purposes. The active principal, atropine, is a potent cholinergic antagonist. As such, it blocks contraction of the radial muscle fibers in the pupil and so causes dilation. Because pupil dilation is a sign of sexual interest in their companion, women would ingest belladonna "of an evening" to enhance their allure. Atropine is still used to dilate the pupil for ophthalmoscopic examination.

d. Police are trained to pay close attention to autonomic function in suspects. Pupil diameter responds to certain illicit drugs. Amphetamine and cocaine are sympathomimetic and dilate the pupils. Heroin and other opiates activate the PS central pathway and cause constriction.

e. Students need to become more effective learners and so should look carefully at Figure 16–13 for guidance. Point to the "repetition loop" acting on the hippocampus. This keeps new information "alive" for long-term storage. If not stored, it gets dumped out the bottom. The hippocampus is the engine that drives long-term memory. It pushes memories into medium- and long-term storage circuits.

5. Common Student Misconceptions/Problems

a. All the adrenergic receptors are G proteins, but only beta-1 receptors activate adenyl cyclase and lead to a rise in cAMP.

b. Try to clarify that, because of autonomic tone, a drug can produce the action of one division by inhibiting the activity of the other. Thus, a "sympatho*mime*tic" drug—a drug that *mimics* sympathetic activation—might be an adrenergic agonist *or* a cholinergic antagonist.

c. The short autonomic reflex involving only peripheral neurons appears to violate the model of the neural reflex taught earlier. In this case, the integrating neuron is not in the CNS but the short reflex still is

   sensory neuron → integrating neuron → motor neuron

   This means that the wiring pattern of autonomic ganglia is far more complex than shown in Figures 16–5 and 16–8. Figure 16–11 clearly shows the minimum wiring needed.

6. Vocabulary Aids

a. Students are sometimes puzzled by the use of the term "*adren*ergic" to describe synapses that release norepinephrine or epinephrine. Explain that these compounds were once known as nor*adren*aline and *adren*aline. Thus, they are named in the same way as *cholin*ergic synapses, wherein acetyl*choline* is the neurotransmitter.

b. *Para*sympathetic should be compared to *para*llel. One meaning of "para-" is "alongside of." The sympathetic and parasympathetic nervous systems work alongside of each other.

c. A "ramus" is a branch or projecting part (recall the ramus of the mandible).

CHAPTER 17

The Special Senses

Introduction

It is our somatosensory system that provides a picture of our body but it is the special senses that most impact our consciousness, inform us about our surroundings, and alert us to threats. One way of getting students to think about the significance of the special senses is to ask which one they think would be the most disabling to lose. Most will say vision. They are often surprised to learn that persons who are both blind and deaf usually choose hearing. To close the discussion, suggest that vestibular senses are the most important. Without them, standing, sitting, even crawling would be almost impossible. Ask them to imagine what it would be like to live with a constant sense of falling or spinning. Suggest a research project about *Meniere disease.* We live in a gravity-burdened three-dimensional world in which a reliable sense of balance is crucial.

Instructional Goals/Learning Outcomes

1. Describe the sensory organs of smell, trace the olfactory pathways to their destinations in the brain, and explain the physiological basis of olfactory discrimination.
2. Describe the sensory organs of taste, trace the gustatory pathways to their destinations in the brain, and explain the physiological basis of gustatory discrimination.
3. Identify the internal and accessory structures of the eye, and explain the functions of each.
4. Explain color and depth perception, describe how light stimulates the production of nerve impulses, and trace the visual pathways to their destinations in the brain.
5. Describe the structures of the external, middle, and inner ear, explain their roles in equilibrium and hearing, and trace the pathways for equilibrium and hearing to their destinations in the brain.

Teaching Strategies

1. Lecture and Lab Ideas

a. Olfactory receptors represent one of the few examples of bipolar neurons. The retina is another location for this rare class of neurons.

b. Even more rare are anaxonic neurons. The only example of them in vertebrate nervous systems is the granule cells in the olfactory bulbs.

c. Clarify that new tastes have been added to the traditionally recognized salt, sour, bitter, sweet. The gustatory sensitivities now include *umami* (savory, in Japanese), triggered by the tastant *glutamate*, water, and astringency. The umami receptor is a G-coupled glutamate receptor similar to synaptic receptors. Glutamate occurs in roasted meats and stews. It is sold as the flavor enhancer, *MSG*. Biophysical, neurophysiological, and behavioral evidence support a distinct *water taste* and the taste of *tannins*, the plant material that puts the "pucker" in strong tea. Some afferents in NV respond to capsaicin, the compound that makes peppers hot. Some consider this sense gustatory, others nociceptive.

d. Olfactory experience often leads to particularly powerful and persistent memories. The olfactory cortex is where smells are experienced and their memories stored; it is part of the "old cortex," that is, the one we inherited from simpler vertebrates such as fish and reptiles. To emphasize the potency and primitiveness of olfactory memory, remind your students of the olfactory memory that guides anadromous fishes such as salmon and eel back to their home stream after many years in the pelagic ocean.

e. Reflex control of pupillary diameter relies on afferents that skip the thalamus and head straight to the midbrain [Figure 17–19].

f. Observe that the retina, unlike the other tunics, covers only the posterior two-thirds of the eye.

g. Since the photoreceptors are at the base of the retina, light must pass through all the other neural and vascular layers before it reaches the photoreceptors. Visual information is relayed and processed as it passes from the receptors up through the neural retina to ganglion cells on the retinal surface.

h. Have students trace aqueous humor from its production in the ciliary body to reabsorption into the venous system.

i. The cornea, the aqueous humor, the lens, and the vitreous body all refract light as it passes through the eye, but the lens is the only structure capable of changing its shape to control the focal distance and influence the focal point.

j. Use Figure 17–19 to clarify that the central projection of the retina is ipsilateral for the temporal field and contralateral (via the optic chiasm) for the nasal field. Thus, the left visual cortex has information from both eyes for the right visual field, and vice versa.

k. It is interesting that, except for vision, all the receptors of the special senses have "cilia" (microvilli).

l. When presenting Figure 17–14, comment that the detail we possess about photoreception is a tribute to the central role that vision plays in human life.

2. Analogies

a. The eyelids (*palpebrae*) are like windshield wipers on your car. They spread tears (wiper fluid) over the cornea and sweep away any accumulated dust and dirt.

b. Compare the wave that travels along the basilar membrane from the oval window toward the helicotrema to moving a bullwhip by whipping the thin *tip end* back and forth. Only the very low frequency waves would make it very far from the thin end toward the thick butt end of the whip.

c. The olfactory ability of a dog compared to a man is like the visual ability of a person with 20,000,000/20 vision.

d. The "eye as camera" analogy still works with a digital camera (with a chip replacing the film for retina) because the lens is still apparent, but as photography moves to the cellular telephone, the analogy will no longer be apparent. "My eye is like my cell phone? Huh?" [Of course, the cell phone camera has a "lens"—a high-performance NSA-developed pinhole lens.]

3. Demonstrations

a. Illustrate how the conjunctiva forms a continuous epithelial membrane from the inner surface of the upper eyelid, along the surface of the eye, to the inner surface of the lower eyelid.

b. To demonstrate how the eye focuses on close or distant objects, ask the students to hold their index finger about 12 inches in front of their nose. If they focus on their finger, the front of the room will appear blurred. If they focus on the front of the room, their finger will blur. That is, the lens can focus only at one distance at a time.

c. During the above demonstration, have another student watch the pupillary diameter carefully. If there is enough difference between the near and far points, they should see constriction accompany near focus. This is a sign that outflow from the oculomotor nucleus in CN III innervates both the ciliary muscles of accommodation and the sphincter of the iris.

d. Obtain an Ishihara color-blindness test set (by purchase or by download). Ask for volunteers who consent to taking a color-blindness test in front of the class. Try to include some color-blind or color-weak (anomalous trichromats) students in the demonstration. Permit the class to observe. They will be surprised by the demonstrable deficiencies of the color blind and color weak. What is the incidence of this "handicap"? What is the gender?

e. Ask them to listen to the shriek their TV makes when first turned on. This is the sound given off by the horizontal sweep circuit, which runs at 15,525 Hz. Your older students will most likely not be able to hear it. Use this observation to motivate a discussion of the place theory of frequency discrimination and how this leads inevitably to high-pitch deafness. Because all frequencies necessarily pass through the region just proximal to the oval window, this is the region that gets the most use and so is the first to wear out. High-frequency sounds are detected here and so high-tone deafness is a common occurrence with advancing years. In addition, exposure to loud sounds accelerates this hearing loss. (Presbyotia—even Lutherans get it!)

f. Point to the puffiness beneath the skin in the superior lateral part of the orbit. Let the class know that here they can see the lacrimal glands under the skin. Does this region become puffier after crying? The greater tissue volume indicates greater tissue activity (elevated blood flow, cell water).

g. A simple paper strip about one inch by six inches with two distinct visual targets (**X** and **O**, for example) about four inches apart can be used to demonstrate the "blind spot." Use Figure 17–4 to show how the presence of the optic disc in the nasal retina (lacking receptors) leads to a field defect ("scotoma") in the temporal field based on image reversal by a lens.

h. Someone appeared on TV performing the "stupid human trick" of taking milk into their mouth and then squirting it out the corner of their eye. Ask the class to explain the anatomy of this stunt.

4. Applications

a. The concept of receptive field from somatosensory physiology can be applied in the visual system. The part of the visual field that excites or inhibits a ganglion cell is its receptive field. Similarly, frequency discrimination depends on the place along the frequency scale as it maps to the basilar membrane.

b. Olfactory receptors represent the largest multigene family in multi-cellular organisms. More than 10,000 unique receptors can be encoded by the genome. The 2004 Nobel Prize in Physiology or Medicine recognized these discoveries.

c. A viral or bacterial infection of the epithelium that lines the palpebrae and covers the anterior sclera is conjunctivitis ("pink eye"). It is often markedly contagious among children because of frequent hand–eye contact.

d. Eye ache from close visual work is a form of muscle cramp (of the ciliary muscle). Looking at a distant object relaxes the ciliary body, relieving the discomfort.

e. Bell palsy is a flaccid paralysis caused by injury to CN VII (facial nerve). Because the palpebrae can't close, these patients must use artificial tears to moisten the eye on the affected side to avoid corneal injury.

f. Ask the students, why, after an eye examination requiring pupillary dilation, are you so sensitive to bright light? Also, why is it so difficult to focus closely, which makes reading very difficult?

g. The chart of letters used to measure visual acuity that gradually decreases in size is called the Snellen Eye Chart. Point out that the ratio used to describe visual acuity as measured this way includes the individual's results in the numerator as compared to the rest of the population in the denominator. Thus, 20/400 vision (the legal threshold for blindness) means a normal person can see at 400 feet what a legally blind person can see at 20 feet.

h. The pharyngotympanic (or, Eustachian) tubes may be too narrow to drain accumulated fluid effectively in very young children. This leads often to the painful condition otitis media; these children will often "outgrow" these problems as the tube develops a larger lumen.

i. A spinning dancer's ability to resist the dizziness from turning is based partly on use of the "spotting" technique (where they fix their gaze on one point except for brief interruptions as they spin) and on experience and practice. Novice dancers at first experience dizziness but learn over time not to become dizzy.

j. Arthritis of the auditory ossicles or scarring secondary to an infection in the middle ear can lead to conduction deafness. In some cases, microsurgery to free up the joints can restore or improve hearing. In others, reconstructive surgery using man-made materials can be effective.

k. Horner's syndrome (mydriasis, ptosis) results from damage to the sympathetic innervation of the eye possibly from a neck injury that involves the superior cervical ganglion. As a result, constriction of the pupil occurs, produced by the intact parasympathetic innervation, unopposed by sympathetic dilator fibers.

l. Dark adaptation can be preserved when you enter the light by wearing red glasses. Rods are insensitive to red light (see absorption spectrum in Figure 17–16). Radar operators and darkroom workers use this applied physiology trick.

m. Lack of a consensual pupillary response (illuminate one eye, the contralateral pupil contracts) is a sign of midbrain injury used in a neurological assessment.

5. Common Student Misconceptions/Problems

a. Students may not realize that the gradual disappearance of an olfactory perception does not result from dissipation of the odorant, but rather from peripheral and central adaptation.

b. Students may think that the lingual papillae visible on the tongue are the taste buds. Clarify that the taste buds are recessed in pits around the papillae.

c. Students may not realize that the sclera does not stop at the edge of the visible eye but continues as a sphere that covers the entire eye. A three-dimensional eye model is helpful on this.

d. Students are often surprised to discover that tears form at the superior, lateral portion of the orbit; most think they form at the medial canthus. Emphasize that tear production at the superior fornix enables the tears to flow downward and medially, moistening the exposed conjunctiva.

e. Students have difficulty uniting the sense of hearing and the sense of balance within the inner ear. It may be helpful if you point out that perilymph fills most of the interior of both components.

f. Most diagrams of the eye show a longitudinal sectional view. Students sometimes don't recognize that the ciliary body, ciliary processes, and suspensory ligaments form a circle; they are usually familiar with the iris and its pupil, both visible.

g. The process of accommodation can be confusing to students, because of lack of understanding of optics and the mechanics of the lens–ciliary body system. Explain the process in a step-by-step manner as you diagram each step:

   1. A highly curved lens bends light more strongly; this is needed to focus on nearby objects. For distant objects, a flatter lens focuses correctly.
   2. The suspensory ligaments connect the lens to the ciliary muscle all the way around the lens equator. Tensing these ligaments flattens the lens; slackening the ligaments permits the lens to assume its natural curved shape.
   3. The ciliary muscle is a sphincter muscle; when it contracts, its diameter decreases, slackening the suspensory ligaments, and vice versa.
   4. Thus, contraction of the smooth muscle fibers that encircle the lens within the ciliary body slackens the fibers, permitting the lens to round up, and thus focusing nearby objects on the retina.

h. Students sometimes confuse N I with the olfactory tract, expecting a nerve trunk such as N II. Point to the tiny axon bundles that penetrate the cribriform foramina of the ethmoid bone to synapse in the olfactory bulb.

6. Vocabulary Aids

a. Associate the "R" in ***R***od with the "R" in da***R***k.

b. Associate the "C," in ***C***one with the "C" in ***C***olor. Point out that the three types of cones detect red, green, and blue, not the primary colors red, yellow, and blue as one might suspect.

c. When someone is "nearsighted," they can only "sight the near." When someone is "farsighted," they can only "sight the far."

CHAPTER 18

The Endocrine System

Introduction

Hormones always seem to be in the news. "Oral" growth hormone is hawked over the Internet. Can it be the real thing? Anabolic steroids and other performance enhancement drugs, although officially banned in most sports, nonetheless are widely used by both professional and amateur athletes, as well as others. What are the true benefits and risks? Also, melatonin, DHEA, and other potent hormones and prohormones may be purchased over the counter. How do birth control pills affect a woman's health, and do the risks of estrogen and progesterone replacement therapy outweigh the benefits? Finally, the pandemic of obesity and consequent type II diabetes mellitus, a grave endocrinopathy, is recognized as a leading societal and medical problem in several industrialized nations, including the United States. For these reasons, stirring interest in the topic of hormones and the endocrine glands should not be difficult.

Instructional Goals/Learning Outcomes

1. Explain the importance of intercellular communication, describe the mechanisms involved, and compare the modes of intercellular communication that occur in the endocrine and nervous systems.
2. Compare the cellular components of the endocrine system with those of other systems, contrast the major structural classes of hormones, and explain the general mechanisms of hormonal action on target organs.
3. Describe the location, hormones, and functions of the pituitary gland, and discuss the effects of abnormal pituitary hormone production.
4. Describe the location, hormones, and functions of the thyroid gland, and discuss the effects of abnormal thyroid hormone production.
5. Describe the location, hormone, and functions of the parathyroid glands, and discuss the effects of abnormal parathyroid hormone production.
6. Describe the location, structure, hormones, and general functions of the suprarenal glands, and discuss the effects of abnormal suprarenal hormone production.
7. Describe the location of the pineal gland, and discuss the functions of the hormone it produces.
8. Describe the location, structure, hormones, and functions of the pancreas, and discuss the effects of abnormal pancreatic hormone production.

9. Describe the functions of the hormones produced by the kidneys, heart, thymus, testes, ovaries, and adipose tissue.
10. Explain how hormones interact to produce coordinated physiological responses and influence behavior, describe the role of hormones in the genera! adaptation syndrome, and discuss how aging affects hormone production.
11. Give examples of interactions between the endocrine system and each of the other organ systems.

Teaching Strategies

1. Lecture and Lab Ideas

a. Stress the distinctly different embryological origins of the *adeno*hypophysis and the *neuro*hypophysis. This will help the students differentiate between the two and connects directly to adult function. The adenohypophysis begins as an evagination of the roof of the pharynx (Rathke's pouch), and thus has a distinctly epithelial (glandular) origin. The neurohypophysis arises as unmyelinated axons extend downward from neurosecretory neurons within the hypothalamus, and so bypass the blood–brain barrier.

b. Given its anatomical location, it is not surprising that the pituitary is controlled directly by hormones released from neurosecretory axons with somas in the hypothalamus.

c. The hypophyseal portal system is the first portal system your class will have encountered if you teach the GI tract later, and have only a limited knowledge of vascular anatomy. To provide a simple model of blood flow, begin with this simple scheme on the board:

Artery → Capillaries (in Organ) → Vein

Then write this scheme below it:

Artery → Capillaries (in Organ 1) → **Portal Vein**
→ Capillaries (in Organ 2) → Vein

d. First, rely on Figure 18–8(a) to present the general pattern of negative feedback control of secretion of four different hormones by the adenohypophysis. Don't forget to point to the Key, and illustrate what it means. Then use panel (b) to walk the class through two alternative feedback schemes.

e. Before explaining the function of ADH, ask the students to describe the effect of a diuretic. Most have heard of this term, but if they are puzzled, ask them to explain to you the effect of taking a "water pill." Once it is established that a diuretic makes a person pee a lot, that is, to get rid of water in the urine, ask them to imagine what an *anti*-diuretic would do. This will lead them naturally to understanding the action of ADH in conserving body water stores by reducing water loss in the urine.

f. The process of synthesis, storage, and release of thyroid hormone, as pictured in Figure 18–11(a), is more detailed than in some A&P courses. However, this offers the student a chance to apply in a coherent way their knowledge of cell biology learned earlier (e.g., receptors, second messengers, ion pumps; protein synthesis in the RER, packaging in the Golgi, exocytosis and endocytosis, lysosomes).

g. Encourage the students to organize their thinking about the actions of epinephrine around a "fight or flight" sympathetic activation. This approach applies the principle that *meaning* fullness for the learner makes it far easier to learn and recall information compared to rote memorization.

h. Point out that glucose would diffuse into cells without active transport. Insulin opens glucose channels in its target cells and glucose enters by facilitated diffusion. Ask the class, "What is the significance that glucose uptake by CNS, kidney, and digestive tract cells is insulin-independent?"

2. Analogies

a. A hormone binding to a receptor is like a phone order to a factory, and the G protein–enzyme complex is like a machine inside the factory cranking out what was ordered, the second messengers.

b. A hormone binding to an extracellular receptor is like ringing the door of a locked house and shouting instructions inside to butlers and maids who bustle about doing your bidding.

c. A hormone fits its receptors just as a key fits its lock and opens it. A hormone antagonist is like a bad key that will go in a lock but can't open it. While stuck in the lock, the good key (the agonist) can't open it.

d. The thyroid follicle is like a vitamin E gel cap. The thin, soft, outside layer is the simple cuboidal epithelium, and the viscous material inside is the colloid.

e. The endocrine system is a powerful amplifier system. As an illustration, a few picograms of TRH leads to release of a few nanograms of TSH, which in turn stimulates the release of a few micrograms of thyroid hormone, which leads to the generation of thousands of excess calories of heat.

3. Demonstrations

a. Point out the word *renal* within the word suprarenal. Then associate the suprarenal glands with the superior pole of the kidney. Showing them Figure 18–14(a) will help solidify this location in their minds. Emphasize that the structural and functional differences seen between the suprarenal medulla and cortex are a result of their different embryological origins. Tell them other glands also have compound embryological origins; the adenohypophysis and neurohypophysis of the pituitary differ in this way, and the C cell precursors (neural crest) migrate into the primordial thyroid (endoderm) during organogenesis.

b. A powerful, attention-getting device is to have, sitting on the podium or table next to you, a urine specimen cup, containing apple juice. Say that, "Today urinalysis of samples such as this (hold up the cup) is done by chemical analyzers. But in earlier times, a diagnosis of *diabetes mellitus*, which means 'honeyed, free-flowing urine,' was made by a physician who tasted the urine." This never fails to gross out the students. As they express their disgust, say, "I don't know. Perhaps it isn't that bad." Then pick up the cup and take a sip. State the three P's of diabetes (*polyphagia*, *polyuria*, and *polydipsia*), taking a sip after each, like a benediction. They will not soon forget this lesson. Then say, "This sample is very sweet-tasting. Should be, it's apple juice."

c. Clearly link cause with effect: That polyuria follows osmotically from hyperglycemia, that polyphagia follows from cell starvation and calorie loss in the urine (glucosuria), and that polydipsia follows from polyuria.

4. Applications

a. For many lactating mothers, just to hear the cry of their (or another's) baby triggers the milk letdown reflex. Women in the class who have breastfed will emphatically confirm this application of the neuroendocrine reflex (auditory stimulus → neural integration → oxytocin secretion → milk ejection).

b. Excess water ingestion leads to diuresis and alcohol suppresses ADH release. Is it any wonder that drinking beer makes some people say "Beer runs right through me"?

c. Analogs of releasing hormones, such as GnRH and GH–RH, can be used paradoxically to *block* release of their targets. The block actually follows a brief period of hypersecretion from the pituitary, followed by long-term suppression due to exhaustion of gonadotropes (to reduce sex hormones) or somatotropes (to treat acromegaly) coupled with receptor down-regulation of target cells in the adenohypophysis.

d. Glucocorticoids such as prednisone suppress ACTH and CRH secretion causing the zona fasciculata to involute. To avoid circulatory collapse when the drug is stopped, prednisone must be withdrawn gradually, permitting the patient's pituitary–suprarenal axis to recover. This is called a "prednisone taper."

e. The most reliable and cheapest source of dietary iodine is iodized salt. It has been estimated that 10 cents worth of iodized salt can prevent a mother from giving birth to an infant suffering from *cretinism* and other developmental anomalies.

5. Common Student Misconceptions/Problems

a. Students are sometimes surprised to discover that FSH, LH, prolactin, and oxytocin are present in both females and males, because these hormones play such a visible role in female reproduction. Let them know that there are many similarities between the male and female reproductive systems. The role of oxytocin in the male is poorly understood but blood levels rise during sexual arousal and peak during orgasm in both males and females.

b. To maintain homeostasis requires constant endocrine communication among the body's cells. This extends well beyond the classically recognized endocrine glands to include almost every organ and cell type in the body. Often students are surprised to see kidneys, thymus, heart, stomach, duodenum, and fat listed as hormone producers. One could add skeletal muscle, epidermis, uterus, placenta, bone marrow, blood vessels, efferent axons, and likely others.

c. It may be difficult for students to see how cells in a diabetic patient can be starved for glucose despite bathing in excessive glucose. This situation is analogous to axons at rest. Although they are soaking in a high-sodium solution, sodium ions don't flood across the membrane until the voltage-gated sodium channels open.

d. Take plenty of time working through the easily misunderstood signaling mechanisms in Figure 18–3. The IP_3 cascade is complex but important.

e. Encourage the students to review the appropriate parts of Chapters 2 and 3 as needed for help with the chemistry and cell biology of the endocrine system.

6. Vocabulary Aids

a. Associate the "C" in **C**ushing with the "C" in excessive **C**ortisol. Associate the "A" in ***A***ddison with excessive ***A***CTH, due to the lack of negative feedback suppression by cortisol.

b. Students will often confuse "glucagon" and "glycogen." First, warn them of this pitfall. Point out that "glyco*gen*" literally means, *gen*erates sugar; associate gluca*gon* with Penta*gon*, a command building, just as glucagon is a command molecule.

c. The point of ***IN***sulin is to get glucose ***IN***to the cell.

d. Androgen comes from "andro-," the combining form for "male," and "-gen," the combining form for "produce." "Android," an artificial human being, may help them remember this origin. Androgens produce "male" secondary sexual characteristics in *both* males and females.

e. Somato*tropin*, somato*medins*, and somato*statin* sound alike to the naïve. Point out, first, that they all begin with the combining form "somato-," which refers to the body. The suffixes give the meaning: "-tropin" refers to turning toward; "-medin" refers to mediate; "-statin" refers to stopping.

CHAPTER 19

Blood

Introduction

Among blood's many functions, oxygen transport is the most acutely important. Because fresh blood may not be available in a trauma situation such as combat or civil emergency and because of the small but real risk of disease transmission by contaminated blood, there is ongoing interest in developing blood substitutes with oxygen-carrying capabilities equivalent to whole blood. Nucci and Abuchowski described the background of this research effort in a February 1998 *Scientific American* article. Most approaches are based on hemoglobin but synthetic materials such as perfluorocarbons are possibilities as well. These synthetic fluids carry enough oxygen in solution that an organism can actually breathe it! A hemoglobin derivative called PolyHeme® underwent clinical trials with gravely ill trauma patients. It was administered in the ambulance where whole blood was unavailable on the way to a trauma unit. Although initially promising, this blood substitute did no better than saline at keeping patients alive, and so the quest for an improved artificial blood substitute continues.

Instructional Goals/Learning Outcomes

1. Describe the components and major functions of blood, identify blood collection sites, and list the physical characteristics of blood.
2. Specify the composition and functions of plasma.
3. List the characteristics and functions of red blood cells, describe the structure and functions of hemoglobin, describe how red blood cell components are recycled, and explain erythropoiesis.
4. Explain the importance of blood typing, and the basis for ABO and Rh incompatibilities.
5. Categorize white blood cell types based on their structures and functions, and discuss the factors that regulate the production of each type.
6. Describe the structure, function, and production of platelets.
7. Discuss the mechanisms that control blood loss after an injury, and describe the reaction sequences responsible for blood clotting.

Teaching Strategies

1. Lecture and Lab Ideas

a. Recall for the students that blood is a fluid connective tissue. That is, blood nicely fits the cardinal properties of connective tissue, a population of cells—the RBCs and WBCs—scattered within an acellular matrix—the plasma. There are striking differences between blood and the other connective tissues, however: the cells were not formed within the connective tissue, and the matrix was not produced by the blood cells but by many other cells past which the blood flows. A good thought exercise is to ask the students to list organs, cells, and processes that impact the composition of the blood. This can serve as a unifying theme for organismic physiology.

b. Students need to learn the clinical units used to report blood test results. Often only the number value is spoken of (e.g., a blood glucose of "90" is shorthand for "90 mg/dL," the units being understood. Refer students to Table 19–3 for the abundance of formed elements, expressed as number/μL. Require knowing the normal ranges for the principal blood cells along with hematocrit normal range, hemoglobin content, and the main electrolytes.

c. The hemostasis process presents two good examples of *positive* feedback at work. The platelet phase shows a rapid acceleration as the activated platelets release factors that recruit additional platelets. The coagulation cascade similarly grows explosively because each step converts a proenzyme to an enzyme. Each enzyme can catalyze many cycles each, and so a rapid amplification occurs. This is not unlike the cascade effect in second-messenger systems.

d. Summarize hematopoiesis (Figure 19–11) this way: stem cells first commit to become either myeloid or lymphoid cells. Myeloid cells then choose among four main lines of differentiation (RBCs, megakaryocytes, granulocytes, monocytes), while lymphoid cells mature to lymphocytes.

e. When describing the structure of hemoglobin (Figure 19–3) point out that interactions between its four subunits underlies the physiologically adaptive oxygen-binding behavior. Also, the globular shape helps enable highly concentrated solutions.

f. Heme breakdown products are what makes urine yellow and feces brown.

2. Analogies

a. The bloodstream is like a river flowing past a city. The people in the city (the cells) pull out of the river what they need (oxygen, nutrients) and dump into it their wastes (carbon dioxide, urea) for downstream processing by the lungs and kidneys.

b. Erythrocytes pass single file through the capillaries, like ticket holders pass the ticket taker at a movie theater; the ticket they hand off is like the oxygen the erythrocytes unload.

c. Mature erythrocytes are enucleated and are shaped like a cookie with a thumb indentation on both sides.

d. The clotting of blood is like the hardening of epoxy glue. A liquid when catalyzed that reacts to assemble polymers from subunits and convert to a solid.

e. Neutrophils defending the host are like honeybees defending the hive. Both die in the process but the host (hive) lives on.

3. Demonstrations

a. Students can demonstrate WBCs moving single file through a capillary plexus by looking at a blue sky. (Blue light is absorbed by RBCs.) They should be able to appreciate the shuffle of bright objects—transparent WBCs traveling amidst the opaque RBCs—within retinal capillaries, often pulsing in synchrony with the heartbeat. They are seeing light passing through WBCs to the photoreceptor layer in the retina below.

b. A fingertip blood sample can be quickly separated into plasma and packed red cells in a microhematocrit centrifuge. Students are always impressed when you show them the tube, divided into two distinct parts. The concept of hematocrit becomes immediately clear. Ask the class why the RBCs are denser than plasma (higher protein and iron content).

4. Applications

a. The plasma proteins play the main role of providing the "colloid" osmotic pressure that sucks fluid back into capillaries from the interstitial space. Inadequate levels, due to liver disease or malnourishment, result in ascites, accumulation of fluid in the peritoneal cavity.

b. A moderate elevation of WBCs (leukocytosis) is often a sign of infection. Very high counts may indicate leukemia.

c. Individuals suffering from alcoholism often have coagulation problems due to liver injury that reduces the levels of coagulation proteins made in hepatocytes. Vitamin deficiencies from poor nutrition also contributes to coagulopathies.

d. Individuals scheduled for surgery are instructed not to take aspirin during the preceding week. Aspirin blocks the enzymes that initiate prostaglandin synthesis and so interfere with platelet aggregation. This leads to prolonged bleeding times.

e. Figure 19–9 does a good job of sorting out the complexities of Rh and pregnancy. Ask the class about their experience with Rhogam (anti-Rh pooled immune globulin).

f. To prevent clotting of isolated donor blood or diagnostic samples, compounds are added (e.g., citrate or EDTA) that complex or *chelate* calcium ion, denying coagulation an essential cofactor.

5. Common Student Misconceptions/Problems

a. Ask the students why the kidney is the logical organ to control the release of erythropoietin. This fact is probably puzzling until you let them know that the kidney receives almost 25 percent of cardiac output and is profoundly dependent on good blood flow to function and remain healthy. Point out that the kidney secretes erythropoietin in response to decreased levels of oxygen, not to decreased levels of RBCs. That is, the kidney is measuring a chemical property and NOT doing a blood count!

b. Blood typing is often confusing to students. They often expect that an exact match between donor and host is required and may therefore find it difficult to see that the Type AB individual can accept Type A or Type B or Type O blood. Point out that compatibility between donor and host is based on the risks that agglutinins (antibodies) in the host plasma pose to the donor red cells.

c. The *intrinsic* and *extrinsic* pathways of coagulation are named with reference to the blood. The intrinsic pathway, the ability of isolated blood to clot on its own, is triggered by platelets, an intrinsic blood component. The extrinsic pathway is triggered by tissue factors, that is, factors that lie outside the blood.

d. The term "formed elements" leads many students to think it means compounds such as water, proteins, etc. Let them know that "elements" is used in an old-fashioned non-chemical sense and "formed" is used in contrast with formless fluids such as plasma.

6. Vocabulary Aids

a. The MONocyte is the MONster cell that gobbles up debris and pathogens.

b. Megakaryocytes, the marrow cells that pinch off their cytoplasm to form platelets, are giant cells with a large ("mega-") multilobed nucleus ("karyo-").

c. *Eos*in, the reddish-pink dye used in Wright's blood stain, is named for *Eos*, the Greek goddess of the dawn.

d. Baso*phils* "love" basic dyes and eosino*phils* "love" eosin. "Phil-" is the combining form for "love."

e. Make sure that they understand the distinction between agglutinin and agglutino*gen*. Point out that the latter term contains the combining form "-gen," which means *gen*erates or gives rise to. That is, an agglutinogen is an antigen that *gen*erates an immune response, the production of an antibody, the agglutinin. Note that *agglutinin* contains the root "glu-." This is because these antibodies can *glue* together red cells that have the matching blood group antigen on their surfaces.

f. *Platelet* refers to "tiny plate," the disk-shaped thrombocyte.

CHAPTER 20

The Heart

Introduction

Classes know that "heart attack" (myocardial infarction) is a leading cause of death in the United States. Most will not know that atherosclerosis is the underlying cause of death from myocardial infarction. And although many know that cholesterol is "bad" for the heart, perhaps the link is not entirely clear. Peter Libby, a Harvard cardiologist, in a May 2002 article in *Scientific American*, presented a new view, now widely accepted, of how atherosclerosis does its deadly damage. Rather than being a mere plumbing problem caused by plaque on the walls plugging the "pipes," the ischemic injury is provoked suddenly by a plaque rupturing through the endothelium to trigger thrombus formation. Remind the class of the positive feedback nature of clot formation learned in Chapter 19. Blood clots can kill quickly by blocking blood flow in a critical organ such as the heart or brain.

Instructional Goals/Learning Outcomes

1. Describe the anatomy of the heart, including vascular supply and pericardium structure, and trace the flow of blood through the heart, identifying the major blood vessels, chambers, and heart valves.
2. Explain the events of an action potential in cardiac muscle, indicate the importance of calcium ions to the contractile process, describe the conducting system of the heart, and identify the electrical events associated with a normal electrocardiogram.
3. Explain the events of the cardiac cycle, including atrial and ventricular systole and diastole, and relate the heart sounds to specific events in the cycle.
4. Define cardiac output, describe the factors that influence heart rate and stroke volume, and explain how adjustments in stroke volume and cardiac output are coordinated at different levels of physical activity.

Teaching Strategies

1. Lecture and Lab Ideas

a. When you begin to talk about the heart, briefly remind the students of relevant information they should recall from earlier chapters. The heart is composed of cardiac muscle cells, a striated involuntary muscle (Chapter 4). Remind them about the basics of muscle activation and contraction (Chapter 10), and ask them to recall the effects of autonomic inputs on heart performance (Chapter 16).

b. Have the students draw a box divided into quadrants; ask them to label each quadrant with a chamber and its functions. Next, add lungs and body; now add arrows to indicate flow. Voila, two flow circuits!

Right Atrium Receives blood from body Sends blood to R. Ventricle	**Left Atrium** Receives blood from lungs Sends blood to L. Ventricle
Right Ventricle Receives blood from R. Atrium Sends blood to lungs	**Left Ventricle** Receives blood from L. Atrium Sends blood to body

c. Point out that clinical evaluation of the heart begins with two basic assessments; ask them what they are (heart rate and auscultation). Coax them with pantomime if need be. Ask them how these tests are conducted, what physiology and anatomy they depend upon, and what information they might yield. Ask if any have had an ECG, and what information it can provide. Assure them that during study of this chapter, new insights about these familiar tests will be gained.

d. State explicitly that valves are NOT active movers, made of muscle, but rather are passive movers, made mostly of connective tissue, whose actions are driven by pressure gradients and flow. Forward flow opens them and reverse flow closes them. If a valve doesn't close properly (valvular incompetence), regurgitation compromises cardiac output, adds a volume overload, and thus raises the workload on the heart. Remember, Work = Pressure × Volume.

e. Have the students trace the flow of blood through the heart in proper sequence from venae cavae to the aorta. Make sure that they name the valves, outflow vessels, lung circulation, and veins. Encourage them to refer to the simple drawing they should have made above, along with Figure 20–6(a), to become completely comfortable and familiar with this, the most basic concept in cardiac physiology.

f. Clarify that the coronary circulation follows the standard plan for systemic blood flow: arteries branch from the aorta to supply capillary networks in the organ following which veins carry the spent blood back to the right atrium. Two features are peculiar to the heart's blood supply: The main vessels are on the surface of the organ (buried under a layer of protective epicardial fat) and the venous return is to the coronary sinus rather than the vena cava.

g. Begin discussion of the initiation of the heartbeat by presenting the dramatic fact that the heart will continue to beat when cut out of the body and placed in a dish of physiological saline. This proves that the heart is self-exciting. Stress, however, that the autonomic nervous system constantly adjusts the rate and strength of the heartbeat through parasympathetic and sympathetic actions on pacemaker tissue and the myocardium itself.

h. Point out that the ventricles never completely empty. Typically, they eject about 50 percent of the blood present at the beginning of systole (the EDV); the volume remaining at the end of systole (the ESV) is directly related to the stroke volume. Write the equation on the board:

$$SV = EDV - ESV$$

i. Cardiac output measures the amount of blood pumped out of the heart averaged over a minute. CO is the essential topic in cardiodynamics. Write the CO equation on the board:

$$CO = SV \times HR$$

Ask the students, “If SV increases and HR stays the same, what will happen to CO?” “If HR decreases, what will happen to CO?” Make sure they understand that this is a multiplicative relationship. Indicate that you are first going to discuss all the factors that can make SV increase or decrease, and then you will discuss the factors that make HR increase or decrease.

j. Rewrite the CO equation on the board, but this time include the quantities that determine SV:

$$CO = SV \times HR\text{, and}$$
$$SV = EDV - ESV\text{, thus}$$
$$CO = (EDV - ESV) \times HR$$

Again, ask the students to use the last equation to puzzle out the answers to questions such as, “If EDV increases, what will happen to CO?” “Decreases?” “If ESV increases, what will happen to CO?” “What if it decreases?” Indicate that you are first going to discuss all the factors that can make EDV increase or decrease, and then you will discuss the factors that make ESV increase or decrease. Thus, you will be explaining factors that affect SV.

2. Analogies

a. Ask the students to explain the following sentence from the first paragraph of this chapter: “. . . blood returning to the heart from the systemic circuit must complete the pulmonary circuit before reentering the systemic circuit.” As obvious as this may be to you, it is not always as obvious to the students. An analogy is a milkman covering his route: he carries full bottles to his customers (red cells with oxygen to the cells along the systemic circulation) and picks up the empties (deoxygenated red cells). The empties need to be carried back to the heart (venous return) and then to the lung for refilling (pulmonary circuit) before they can be carried again to the customers on the systemic circuit.

b. As you discuss the two main factors that determine EDV, present the following analogy, “Imagine I am putting water in a bathtub (venous return) and then close the tap (close the AV valve). How much water will be in the tub?” Prompt the students to say that it depends upon (1) the amount of time you had the tap on, and (2) how fast the water was flowing in. Make an explicit analogy to EDV being affected by the duration of ventricular diastole and the rate of venous return. Ask them what would happen to EDV if the ventricular wall were not fully relaxed? This is the basis of diastolic heart failure. Low EDV then leads to low SV and thus reduced cardiac output. High heart rate also lowers EDV because of reduced filling time.

c. A heart valve permits flow in only one direction. This is like the stop strip that ensures one-way traffic at auto rental locations. It’s easy and safe one way, but shreds tires the other!

d. Fluid engineers would characterize heart valves as "check valves," that is, valves that ensure one-way flow in hydraulic circuits. For example, a check valve is installed beween the municipal water supply and the end user.

3. Demonstrations

a. Have them calculate total blood volume for a typical adult human based on the values in Chapter 19. Then, repeat the calculation of cardiac output based on the values in Chapter 20. Do they see that the heart could effectively empty the vessels of blood in one minute? Would it? Or would cardiac output fall as blood was squirted out? [Yes, it would. This is hypovolemic shock brought on by reduced venous return.]

b. Explain the striking difference between the thickness of the right and left ventricular walls [see Figure 20–6 (a) and (c)] in terms of the work of pressurizing a fluid:

$$\text{Work} = \text{Pressure} \times \text{Volume}$$

The left ventricular peak systolic pressure is 4 to 5 times higher than the right, and so, because the volumes are equal, the left ventricle does 4 to 5 times as much work, and so grows to be four times thicker to match the load. This is a form of exercise hypertrophy expressed by cardiac muscle. Ideally, a preserved animal heart will be available to let them see the functional anatomy for themselves, the best way to learn.

c. State first in words and then with an equation that "venous return equals cardiac output" (because the heart can't make blood but only pump what comes back to it from the periphery). That is,

$$\text{VR} = \text{CO}$$

Thus, factors that reduce VR will automatically reduce CO. Ask them to use this concept to predict what will happen to CO if someone stands up suddenly and gravity pulls blood away from the right atrium into the veins. See if they can see the reflex adjustment by measuring heart rate while sitting and then immediately after standing.

d. Encourage the class individually to count ribs by palpating the position of the second and fifth intercostal spaces bilaterally as landmarks for auscultation of the valves. Fleshy students may need to press hard.

4. Applications

a. Explain that the rapid, weak pulse seen after severe blood loss (hypovolemic shock) is best understood in terms of a causal sequence: Blood loss → reduced venous return → reduced cardiac output → lowered blood pressure → reduced barorecptor output → reflex increase in heart rate. Emphasize the importance of this kind of *diagnostic* or clinical reasoning.

b. To identify which valve is generating a murmur, the physician relies on where on the thorax the sound is heard loudest and when during the cardiac cycle it's heard. The latter information permits judging whether the valve is incompetent or stenotic. Figure 20–18(a) indicates the places where each of the four valves is heard the loudest.

c. The presence of relatively infrequent disturbances of heart rhythm can often be documented if a patient's ECG is continuously recorded with a "Holter monitor" while the patient keeps a diary of daily activities. It is named for its inventor, Norman J. Holter, an American biophysicist.

d. External cardiac massage ("CPR") is effective at maintaining blood pressure and flow under emergency conditions because, by simply compressing the sternum against the pericardial sac, ventricular pressures are raised that close the AV valves and open the semilunar valves, thus forcing blood to the lungs and aorta. Thus, the sophisticated anatomy of the heart enables a simple intervention such as chest compression to sustain blood flow adequate to sustain life. Figure 20–8 nicely illustrates valve form and function.

e. To get a laugh, tell the students that cardioversion by applying a brief powerful current pulse to the thorax through external electrodes does not save a life through the exercise it imposes—jumping off the table. It works by forcing most of the myocytes to generate an action potential simultaneously, and so become refractory about the same time. When excitability returns, they are "captured" by the SA node with a restoration of normal sinus rhythm—ideally.

f. The cross section through the thorax [Figure 20–2(b)] approximates a CT image.

5. Common Student Misconceptions/Problems

a. Students often harbor the mistaken "rule" that arteries carry bright-red, oxygen-rich blood, and veins, the opposite. Use Figure 20–1 to underscore that arteries carry blood away from the heart, and veins toward it. Blood leaves the heart twice and returns twice for every complete circuit!

b. Emphasize that the ECG represents electrical activity in the heart and NOT mechanical (contractile) activity. Students are strongly inclined to equate the ECG signal with contraction of atria and ventricles. The key is to emphasize that depolarization (P and QRS waves) is the *cause* of contraction but the signals are electrical, not mechanical. The arterial pulse gives a clearer picture about stroke volume and thus cardiac output.

c. The duration of a cardiac cycle is the time between the same events in two successive cycles. Thus, cardiac cycle duration (and its inverse, heart rate) is the time between P waves, which includes the time in diastole, and not merely the P–T duration.

d. Clarify for students that the principal heart sounds (S_1 and S_2, or "lubb" and "dupp") mark the beginning and end of systole. The sounds heard through a stethoscope are NOT caused by the leaflets of closing valves slapping together but rather from the shaking of cardiac structures when blood briefly flows backward, closing the valve, and then suddenly stopping.

6. Vocabulary Aids

a. When discussing the *trabeculae carneae* note that "trabecula" (means arch) was a term introduced during the study of spongy bone. Also note that the root element for the word *carn*eae is heard in the words *carn*ivore (meat eater) or chili con *carne* (a meat stew seasoned with chilis). Thus, trabeculae carneae literally mean "fleshy arches."

b. "Intercalate" means "to insert between or among," from inserting days into a calendar. Thus, the *intercalated* discs look like they have been inserted between cardiac myocytes. Here they establish mechanical and electrical continuity through desmosomes and gap junctions. It is through them, visible in the LM, that syncytial behavior by the myocardium is possible.

c. Associate ***A***rtery with ***A***way (from the heart).

d. "Visceral" refers to the epithelial layer on the *viscus*, an organ in the ventral cavity. "Parietal" refers to the layer lining the *wall* of the cavity, from Latin for wall.

e. The first sound "lubb" is softer and more prolonged than the second heart sound "dupp" because the leaflets of the AV valves are larger and consequently close off flow less abruptly than do the semilunar valves, which snap shut.

f. The E*C*G is often called an E*K*G, because "*K*ardio-" is the combining form for "heart" in Dutch and German. Dutch physicist and physician (and 1924 Nobel Laureate) Willem Einthoven developed the first widely successful application of ECG technology. In his honor, we name the triangulation method for measuring the mean cardiac vector, Einthoven's Triangle.

g. "Papillary" refers to a nipple-like muscular extension of the ventricular wall. *Papilla* is Latin for nipple. (Anatomists seem to have a special fondness for that word.)

h. To recall its effect on cardiac performance, point out that the **S**ympathetics **S**peed and **S**trengthen the heartbeat.

i. *Tachy*cardia should remind them of *tach*ometer, a device for measuring engine speed.

j. *Ectopic* will be encountered again in Chapter 29. It derives from the Greek *ecto*, outside of, and *topos*, a place.

CHAPTER 21

Blood Vessels and Circulation

Introduction

The scheme for the circulation of blood (heart → arteries → capillaries → veins → heart) is now universally accepted and taught to all students of anatomy and physiology but this was not always the case. The classical but largely mistaken teachings of the second-century Roman physician and philosopher **Galen** that dominated medical thinking for more than a millennium were eventually replaced by newer insights. Beginning in the Renaissance and continuing to our day, direct experimentation, dissection, and observation trumps authority and confused logic. Andreas **Vesalius** of Brussels (1514–1564) produced in 1543 at the age of 28 Europe's first detailed atlas of the human body. It accurately portrayed the heart, arteries, and veins. In particular Vesalius discovered that the inferior vena cava did not *arise* in the liver, and thus the liver could not be the source of blood, as the galenical system of medicine required. This anatomical knowledge laid the basis for the revolutionary work of William **Harvey**, an English physician, who published *An Anatomical Study of the Motion of the Heart and of the Blood in Animals* in 1628. This monograph rocked the world of science and medicine. In direct opposition to Galen who taught that it was the arteries that pulsed to move the blood and the heart worked like a bellows to suck blood in during diastole, Harvey proved by experiment and observation that blood was pumped from the heart during systole through arteries to the entire body and returned to the heart in veins; that is, *in human beings, the blood circulates.* He asserted that there must exist unseen channels between arteries and veins (the hypothetical *microcirculation*). Capillaries were not directly detected until 1661. Marcello **Malpighi**, an Italian physician and pioneer in microscopic anatomy, saw tiny tubules in teased frog lung and recognized that these were the unseen connections between arteries and veins. So, what today is commonplace knowledge accepted by all was once revolutionary.

Instructional Goals/Learning Outcomes

1. Distinguish among the types of blood vessels based on their structure and function, and describe how and where fluid and dissolved materials enter and leave the cardiovascular system.
2. Explain the mechanisms that regulate blood flow through vessels, describe the factors that influence blood pressure, and discuss the mechanisms that regulate movement of fluids between capillaries and interstitial spaces.

3. Describe the control mechanisms that regulate blood flow and pressure in tissues, and explain how the activities of the cardiac, vasomotor, and respiratory centers are coordinated to control blood flow through the tissues.
4. Explain the cardiovascular system's homeostatic response to exercise and hemorrhaging, and identify the principal blood vessels and the functional characteristics of the special circulation to the brain, heart, and lungs.
5. Describe the three general functional patterns seen in the pulmonary and systemic circuits of the cardiovascular system.
6. Identify the major arteries and veins of the pulmonary circuit.
7. Identify the major arteries and veins of the systemic circuit.
8. Identify the differences between fetal and adult circulation patterns, and describe the changes in the patterns of blood flow that occur at birth.
9. Discuss the effects of aging on the cardiovascular system.
10. Give examples of interactions between the cardiovascular system and each of the other organ systems.

Teaching Strategies

1. Lecture and Lab Ideas

a. As you describe each layer of the blood vessel wall, relate its structure to its functional role. For example, ask the students: "Why is the intima formed from a simple squamous endothelium?" (smooth, thin) "What is the elastic lamina for?" (to drive blood flow during diastole) or "What is the function of smooth muscle in the media?" "What would happen to the diameter of the lumen if it contracted?" "How would that affect the resistance to blood flow?" and "Why is the smooth muscle layer thicker in an artery than in a vein?" Ask, "Why do large blood vessels have their own blood vessels (*vasa vasorum*)?" In discussing the adventitia, note that arteries and veins are among the few organs that lack an external epithelium.

b. Because capillaries are the only blood vessels whose walls are thin enough to support effective diffusional exchange with the tissues, they are the business end of the circulation.

c. When reviewing Summary Table 21–1, Key Terms and Relationships Pertaining to Blood Circulation, address these points clearly:

   - Hydrostatic pressure pushes outward against the walls from inside the blood vessel.
   - Because fluid flows down a pressure gradient (just as molecules diffuse down a concentration gradient), the drop in pressure from aorta to right atrium or from pulmonary trunk to left atrium is what drives flow in the two circuits.
   - Blood pressure refers to the hydrostatic pressure of the blood in the arteries.
   - The "applied force" used in the definition of hydrostatic pressure refers to the action of ventricular myocardium contracting against the blood.

d. Remind the students that plasma proteins provide the colloid osmotic pressure gradient (a role stated earlier in Chapter 19) that pulls fluid into capillaries from the interstitial space as pictured in Figure 21–12.

e. As you use Figure 21–14 to describe the blood pressure control reflex loops, first, beginning at the center, trace out, step-by-step, the effects of a rise above normal blood pressure and then analyze the response to a fall below normal in the same systematic way.

f. Introduce the topic of fetal circulation by asking, "Is it necessary for the fetus to send blood to the lungs for oxygenation?" Point out that the foramen ovale and ductus arteriosus shunt blood from the right side to the left side of the heart, skipping the pulmonary circuit.

g. The contribution of hormonal mechanisms to cardiovascular regulation presents a good opportunity to remind students of the complementary functions of the nervous and endocrine systems.

h. Contrast the effects of angiotensin II with those of atrial natriuretic peptide.

i. Have the students rationalize why patients with hypertension are instructed to drastically reduce sodium intake.

j. Point out that the arteries that branch from the aortic arch are not symmetrical. A brachiocephalic artery on the right, which divides to form the right common carotid and subclavian arteries, is not matched on the left. There, the left common carotid and subclavian arteries branch individually off the arch. Pigs have this pattern too, but other animals, such as the cat, have a different pattern.

k. The subclavian artery continues as the axillary artery, which becomes the brachial artery, which then branches to become the radial and ulnar arteries. Point to the anatomical markers that determine when the name changes. Use Figure 21–29 to illustrate these points.

l. Make these points as you describe the carotid arteries and the blood supply to the brain:

   - The branch point of the common carotid artery into the internal and external carotids is called the carotid bifurcation ("two-pronged").
   - The carotid sinus, the location of the baroreceptors, is located at the branch of the internal carotid artery from the common carotid. The internal carotid serves the brain. What is the functional significance of this location?
   - Pressure applied to the carotid sinus can stimulate baroreceptors, increasing their rate of firing. Ask students to puzzle out the causal sequence that can lead from this input to unconsciousness ("the sleeper hold").
   - To remember that the *internal* carotid feeds the brain while the *external* carotid feeds the face, state that the brain is *internal* while the face is *external*.
   - Use Figure 21–22 to point out that the vertebral arteries ascend through the transverse foramina of the cervical vertebrae. The presence of these foramina distinguishes cervical from thoracic and lumbar vertebrae, as described in Chapter 7.
   - The (anterior) internal carotid arteries directly supply the cerebral arterial circle (circle of Willis); the posterior (vertebral) arteries get there by way of the basilar arteries.

m. The common iliac artery forms a continuous line feeding the femoral, popliteal, and posterior tibial arteries. Anatomical landmarks determine the boundaries of each artery.

n. As an introduction to the hepatic *portal* system, first point out that veins that drain the digestive organs don't drain directly to the inferior vena cava. Rather the mesenteric veins empty first into the hepatic portal vein. This arrangement delivers absorbed foodstuffs undiluted to hepatocytes for storage, just as the hypophyseal portal system delivers undiluted hypothalamic hormones to the adenohypophysis.

2. Analogies

a. The arterial side of the circulation can be compared to a highway system. The large freeway (aorta) carries the heavy traffic. At exits, traffic can divert to lesser highways (distributing arteries) and from there to local roads (small arteries) until they turn into a street that connects to their driveway (arterioles and capillaries). Note that main roads are sometimes called "arteries." Of course, this analogy is too simple: Traffic does not circulate nor do highways adjust their resistance to traffic flow (although traffic lights act like precapillary sphincters).

b. Compare the elastic fibers in the wall of an elastic artery to the rubber in the wall of a balloon. As pressure increases, the balloon inflates, stretching the wall and storing energy. When the balloon is opened, the elastic forces in the wall squeeze the air out.

c. Regional vasoconstriction can shuttle blood from place to place much like a finite military force can be deployed to one hot spot after another.

d. Compare the flow of blood from aorta to vena cava to the flow of water in a river. High in the mountains where the stream arises and where the pressure gradient is steep, water flow is rapid. As the river approaches the coastal plain, where the pressure gradient is flattest, the flow slows to a lazy pace until it finally enters the ocean, from whence it sprang with the rains. The heat of the sun, like the work of the heart, raises the water back to the atmosphere, imparting the potential energy to flow downhill once again.

e. Because only molecules small enough to fit through the pores of a capillary can be filtered out of the blood, the capillary wall acts like a coffee filter. The grounds stay behind and the coffee filters through.

f. The aorta is like a walking cane—the ascending and arch regions form the curved handle and the thoracic and abdominal regions are the staff.

3. Demonstrations

a. Relate pressure, flow, and resistance to sucking on a soda straw or using a garden hose.

b. It might help to demonstrate why a rise in carbon dioxide leads to a fall in pH in terms of the formation and ionization of carbonic acid. This will help them later when respiration and acid–base balance are discussed.

c. It is easy to repeat William Harvey's demonstration that veins contain valves that direct blood flow back to the heart. Choose a well-muscled student with scant subcutaneous fat to make it easier to demonstrate. Directions for performing this investigation can be found online by searching for "William Harvey's Demonstration of the Valves in Arm Veins" at Google.com.

4. Applications

a. The students may have heard LDL referred to as "bad cholesterol." Refer them to the Clinical Note on Arteriosclerosis to gain a better understanding of this disorder and the role of diet in its cause and treatment.

b. Comparison of the strength of the radial pulse to the dorsal pedal pulse provides information about the presence of plaque in the popliteal arteries.

c. Auscultation of the common carotid with a stethoscope may disclose a turbulent sound (called a bruit, pronounced "brew-E"). This is caused by partial occlusion by atherosclerotic plaque. If severe, the risk of stroke is greatly elevated. Surgical removal of the plaque by endarterectomy is then indicated if blockage exceeds 70% or in patients showing clinical signs of impaired blood flow to the brain.

d. A great saphenous vein is commonly sacrificed to serve as graft material for use in coronary artery bypass surgery. Postoperatively, venous return from that leg is badly compromised until a collateral circulation can form many weeks later.

e. In sickle cell anemia, deoxygenated hemoglobin assembles into fibers, forcing the red cells to become rigid. They no longer can fit through capillaries. This denies exchange to the cells. Ultimately, extensive tissue destruction occurs and life span is reduced. This genetic disorder emphasizes the crucial role that capillary exchange plays in body function.

5. Common Student Misconceptions/Problems

a. Although you may define an artery as a vessel that carries blood away from the heart, students have trouble letting go of the misconception that arteries carry oxygenated blood and veins carry deoxygenated blood. Point out that although this is true in the systemic circulation, it is simply not so in the pulmonary or fetal circulations. Emphasize that an artery is defined by the *direction* of blood flow and NOT by the amount of oxygen within the blood.

b. Students who have difficulty visualizing what is meant by a capillary "bed" or plexus will benefit from study of Figure 21–5. Panel (a) shows a tiny network of a few capillaries between an arteriole and venule; panel (b) portrays an actual micrograph of the microcirculation with the vascular elements clearly identified. Impress upon them that all organs are ultimately served by the circulation at this microscopic level.

c. Many students panic when they are confronted with equations, because they do not see them as physical relationships between actual quantities. Talk them through the mathematical relationships as you describe the physiology represented.

d. Because the ventricles are in diastole for a longer period of time than they are in systole, the mean arterial pressure (MAP) cannot be calculated by simply taking the arithmetic mean of systolic and diastolic pressure, (SP + DP)/2. Rather, a formula is needed that gives more weight to the diastolic pressure. The following approximation is widely used (it is mathematically equivalent to the one in the text but easier to remember and more direct in its meaning):

$$MAP = (2DP + SP)/3$$

Thus, for SP = 120 mm Hg and DP = 80 mm Hg, MAP is about 93 mm Hg. Write an example on the board, but have the class do the calculation. Try different values of SP and DP to compare their effects on the mean arterial blood pressure.

e. To make the task of learning the blood vessels manageable be selective about your expectations. To help your students learn the vessels and their names, encourage them to approach the task visually and functionally before trying to memorize the names. Suggest that they first form a mental picture of the principal arteries and veins from a careful consideration of Figure 21–20 and Figure 21–27. Also, pass along these pointers:

- Blood vessels are usually named for location or the organs they serve. For example, "subclavian" means "below the clavicle" and "renal" refers to the kidney.

- Most (but not all!) vessels are paired, left and right. Further, arteries usually lie side-by-side and have the same names as their companion veins (e.g., renal artery and renal vein).
- If a vessel's name includes an adjective describing an anatomical direction, there is usually a nearby vessel with the opposite adjective. Hence, if there is an *external* iliac artery, there will be an *internal* iliac artery, and both an inferior and superior vena cava.
- Practice the vessel pathways by selecting an organ and relating the series of arteries and veins that reach it.

f. Remind students that veins carry blood TOWARD the heart. To trace from the superior vena cava to the brachiocephalic veins is to move "upstream."

6. Vocabulary Aids

a. Enunciate "*arterio*sclerosis" and "*athero*sclerosis" carefully so the students can hear the differences in the words to alert them to the differences in spelling and meaning. Point out that *sclero-* means hard and *osis* refers to a disorder. Arteriosclerosis is the general condition of hardening of the arterial wall, while atherosclerosis is a specific type of arteriosclerosis. *Athero-* means "gruel" in Greek, which alludes to the coarse, friable appearance of *athero*sclerotic plaques.

b. "Vena cava" literally means "hollow vein," referring to its cavelike size.

c. Veins are sometimes referred to as *capacity* vessels, because they are easily expanded to hold blood. Venous compliance is adjusted by sympathetic input to the media, an essential element in the reflex control of the circulation. The veins contain a majority of the blood volume.

d. The different names for the aorta designate the parts of a single vessel as it travels away from the heart, and not the names of distinct vessels. The aorta is such a long vessel that it is useful to describe its parts as ascending arch, descending thoracic, and descending abdominal.

e. The word *celiac* comes from the same root as "coelom," which means "cavity." The celiac trunk is the first branch off the abdominal aorta.

f. The "mesentery" is the fold of the visceral peritoneum that envelops and supports most of the digestive organs. It comes from "mes-," meaning "in the middle of," and "enteron," indicating the intestine.

g. The term *fibular* is now preferred to *peroneal*. This should eliminate confusion between peroneal and *perineal*.

h. Don't confuse the *basilic* artery in the arm with the *basilar* artery at the base of the brain.

i. In a portal system, blood is carried by a *portal* vein to a second capillary bed before returning to the heart. Portal derives from the Latin for *carry*.

j. The umbilical arteries and vein are named based on their relation to the fetal heart, not the maternal heart.

k. *Baro-* is the combining form for "pressure." It is helpful to remember *baro*meter, an instrument for measuring atmospheric *pressure*.

CHAPTER 22

The Lymphoid System and Immunity

Introduction

Students find autoimmune diseases such as lupus erythematosus, multiple sclerosis, Graves' disease (autoimmune hyperthyroidism), and rheumatoid arthritis both interesting and threatening. Because the autoimmune illnesses strike younger women more often than others, it should not be difficult to motivate study of this topic. Alternatively, students are well aware that infection of the immune system (specifically, of macrophages and CD4 cells) by human immunodeficiency virus (HIV) often leads to the collapse of the immune system that culminates in acquired immune deficiency syndrome (AIDS). These issues, along with organ and tissue transplantation, make this topic one of the most vital from the viewpoint of contemporary society and medical care.

Instructional Goals/Learning Outcomes

1. Distinguish between nonspecific and specific defense, and explain the role of lymphocytes in the immune response.
2. Identify the major components of the lymphoid system, describe the structure and functions of each component, and discuss the importance of lymphocytes.
3. List the body's nonspecific defenses, and describe the components, mechanisms, and functions of each.
4. Define specific defenses, identify the forms and properties of immunity, and distinguish between cell-mediated (cellular) immunity and antibody-mediated (humoral) immunity.
5. Discuss the types of T cells and their roles in the immune response, and describe the mechanisms of T cell activation and differentiation.
6. Discuss the mechanisms of B cell activation and differentiation, describe the structure and function of antibodies, and explain the primary and secondary responses to antigen exposure.
7. Describe the development of immunological competence, list and explain examples of immune disorders and allergies, and discuss the effects of stress on immune function.
8. Describe the effects of aging on the lymphoid system and the immune response.
9. Give examples of interactions between the lymphoid system and each of the other organ systems.

Teaching Strategies

1. Lecture and Lab Ideas

a. The ultrafiltration of lymph from capillaries was described in Chapter 21. Remind the students that about 3 to 4 liters/day of lymph are produced. Now focus them on the question of where this large volume of fluid goes. This serves as a good place to begin a study of the lymphoid system, where fluid drainage and host defense are combined.

b. Use Figure 22–12 to clarify the importance of the complement system. Make it clear that the complement system extends the tagging of bacteria with antibodies to cell wall destruction, bacterial lysis, and killing bacteria. The complement proteins work in a cascade to explosively accomplish this. This represents another example of positive feedback, just like blood clotting. Complement also promotes phagocytosis of antibody-tagged bacteria and promotes inflammation.

c. As you discuss the types of acquired immunity, first define the terms *active*, *passive*, *natural*, and *artificial*. Point out that active/passive are used to distinguish whose immune system produced the antibodies; natural/artificial are used to describe the method in which the antigen or antibody entered the body. Organize the material on the board as a table and solicit examples of each from the class:

	Active	**Passive**
Natural	ACTIVE NATURAL e.g., infection by chicken pox virus	PASSIVE NATURAL e.g., maternal antibodies crossing the placenta
Artificial	ACTIVE ARTIFICIAL e.g., immunization against polio with live or killed virus	PASSIVE ARTIFICIAL e.g., antivenom injected against snake venom

d. It is easy to place too much emphasis on the familiar lymphoid organs (nodes, thymus) and give short shrift to the extensive and massive lymphoid tissues (tonsils, adenoids, Peyer patches, appendix, and the large lymphocyte populations in loose connective tissue beneath epithelia). Give the latter their proper recognition. Seventy percent of lymphoid tissue is MALT.

e. The nonspecific defense system is very important if not as “sexy” as the specific immune system. Give it adequate attention in your presentations.

2. Analogies

a. Describe the endothelial cells of the walls of the lymphatic capillaries as flaps that overlap like the scales of a fish, anchored at one end but loose at the other. The loose end opens like a trapdoor, permitting lymph and cells to flow toward the heart but not away.

b. Larger lymphatic vessels are like veins: thin-walled, low-pressure, with valves that direct flow back toward the heart.

c. As you describe the different immune cells and their function, compare them to the different units in the organization of military troops. For example, in the Immune

Corps there are two divisions: nonspecific immunity and specific immunity (because both divisions are fighting the same war, there is often shared action between the divisions). Each division is made up of regiments. The lymphocytes are a regiment of the specific immune division. The T and B cell battalions within each regiment are divided into platoons, such as helper T cells, cytotoxic T cells, and suppressor T cells (NK cells are a special unit).

d. Cytotoxic T cells make physical contact with the target cell. That is, they engage in cell-to-cell combat like two soldiers in hand-to-hand combat. B cells blast out antibodies that reach their targets far away. This is like using poison gas on enemy troops.

e. Students may think of the lymphocytes as fixed in one location. Remind them that they camp out for a while in a node and then move to another. The troops (cells) of each platoon are on constant surveillance, and must frequently change positions as they guard the body against invaders. They travel the highways (blood vessels and lymphatics) to camp in a fort (lymphoid tissues) located elsewhere in the body.

f. The platoons of T cells must reside at "Fort Thymus" while they receive special "training." Once they receive their orders (antigen recognition and T cell activation), they can sally forth against their specific foes.

g. The spleen acts like a combination of quality control station and recycling center, recognizing and destroying rigid or misshapen red cells and then recycling the iron and amino acids they contained.

h. Phagocytes are special battalions of the WBC regiment in the Nonspecific Division. These battalions are made up of platoons of microphages and macrophages. They act as the scouts and are strategically located in areas that are likely to be compromised by the enemy. Microphages are usually short-lived: they are destroyed by a single phagocytic event like kamikaze pilots. Macrophages are larger, so they can survive several encounters with the enemy. Macrophages also help out by cleaning up the debris left from the microphage battles and so deny food to the enemy. Some macrophages are stationed in permanent locations and wait for the invaders to arrive; others are allowed to wander around looking for trouble.

i. Endothelial markers and chemotactic factors are like putting up the flag on a mailbox that ensures a visit by the mail carrier.

j. Specific immunity is the second division of the Immune Corps. Its battalions of B and T cells rely on some of the troops in the first division, but they fight their battles using different weapons, so they need different training. Once they receive their orders (differentiation), they can move around in order to detect infiltrators. There is no military analog of B cells (no "smart" weapons that stick to and destroy their intended victim), but some T cells engage in "hand-to-hand" combat.

k. Compare nonspecific defenses and specific defenses to getting a phone call. Normally, the telephone ring sounds the same irrespective of whom the caller is and we pick up, before knowing the identity of the caller. However, with Caller ID (the specific defense system), we recognize the caller and can respond in a specific way—a much more focused, efficient, and effective process.

l. The network of sinuses within the lymph node can be compared to an elaborate maze.

m. The antibody response to an antigen is like ordering a pair of custom-made shoes. The first order is slow to be filled because a custom form needs to be made to fit your feet. The next time you order, the response is quick. The shoe form is like a memory cell.

n. T helper cells are like translators who convert the message from the antigen-presenting cells of the nonspecific response to the specific immune responses.

3. Demonstrations

a. Different colors of pipe cleaners (from a craft store) can be used to build models of the several categories of immunoglobulins, using a color code for the constant and variable regions.

b. The bump that develops rapidly at the site of a mosquito bite is a familiar example of inflammation. The itch is a sign of histamine release from mast cells. The relief of the itch by compounds such as diphenhydramine that block the histamine receptor demonstrates the role of histamine as an inflammatory mediator.

4. Applications

a. Many adults now in their 60s and older no longer have their palatine tonsils. It was routine to remove the tonsils of children a generation or two ago for tonsillitis. Outcome research showed this lymphoid tissue is better left in place in most cases, and to use antibiotics to treat the infection.

b. The universal vaccination of the world population against smallpox led to the eradication of this dread disease, a first in medical history.

c. CD4 cells (helper T cells) are infected and destroyed by HIV, along with macrophages. This loss accounts for the deadly virulence of this infection.

d. Alpha-, beta-, and gamma-interferons, proteins of the nonspecific defenses, are available from recombinant DNA technology for use as medicines. Alpha-interferon is valuable for treating certain viral infections, beta-interferon benefits patients with multiple sclerosis, and gamma-interferon is approved for a rare but lethal genetic immune disorder.

e. Pooled gamma globulin (immunoglobulin G) isolated from human blood plasma can be injected before or after infection to provide passive immunity against viral infections, such as hepatitis A.

5. Common Student Misconceptions/Problems

a. Most students are surprised to find there is a second network of vessels throughout the body that does not carry blood but is indirectly powered by the heart. Doubtless, they have experienced "swollen glands," but many are unclear about what those glands do and do not know that the nodes are connected by lymphatic vessels. State that node enlargement occurs because of cell hypertrophy and division in response to antigen and cytokine exposure. Bigger, more numerous cells lead to "swollen glands."

b. A network of vessels that is not a circuit (i.e., the vessels carry the lymph from the periphery to the veins) is unfamiliar to the students. Of course, because the lymph is produced by filtration in the tissues, only vessels that carry the lymph back *to* its source (the blood) are needed.

c. Students know from experience that there is pain when an injury becomes inflamed, so they have a hard time believing that the inflammatory response is a GOOD thing. In this process, the pathogen or harmful substance is isolated, destroyed, and healing of the damaged tissue begins. The inflammation process pictured in Figure 22–13 should help clarify the role that inflammation plays in the defense process. Point out the central role played by mast cells and histamine.

d. Students may be puzzled by the idea of self antigens, because they often equate "antigens" with something that is part of a pathogen. Point out that an antigen is simply a chemical flag or marker on a cell's surface. Explain that our cells have a unique profile of protein markers on their surface but that during B and T cell development those that recognize self antigens are destroyed, creating a condition of *tolerance* toward those markers. On the other hand, an organ transplanted to another person would be quickly attacked and destroyed because its cells are recognized as foreign.

6. Vocabulary Aids

a. The *afferent* and *efferent* lymphatics are named in somewhat the same way as the afferent and efferent nerves. Afferent lymphatics carry lymph TO nodes and efferent lymphatics AWAY.

b. Associate the "***T***" in cor***T***ex with ***T*** cells to help learn that T cells are found in the deep cortex of the node.

c. The term *complement* refers to the fact that this system complements the action of antibodies.

d. *Perforin* gets its name because it assembles to punch a leak channel through the plasma membrane. That is, perforin "perforates" the target cell.

e. Students may have an easier time recalling the four cardinal signs of inflammation if you say them in Latin, which rhymes:

   Dolor (pain)
   Rubror (redness)
   Calor (heat)
   Tumor (swelling)

   Classy, if nothing else!

CHAPTER 23

The Respiratory System

Introduction

Lung cancer, emphysema, and asthma—three diseases of the respiratory system—are among the gravest health problems. Although much attention focuses on breast cancer, in fact, lung cancer kills many more women. Lung cancer claims the most male victims as well. Emphysema is a major crippler and legions die a horrible death from it. Its incidence is up 40 percent since 1982. Asthma now troubles about one in eight in the United States, and is fatal to thousands annually. Smoking plays a dominant role in cancer and emphysema and, along with second-hand smoke, contributes to asthma. This chapter provides an ideal circumstance to relate injuries by tobacco smoke to specific components of respiratory function. First, the ciliary system is paralyzed and then denuded by smoking. Smoke triggers the protective cough reflex and bronchoconstriction along both short and long reflex pathways. Chronic inflammation attracts the neutrophils whose elastase and other proteases digest and destroy the elastic fibers in the lung, setting the stage for collapse of alveoli and airways. Finally, the bronchial epithelium, subjected to repeated mutagenic insult, may develop a bronchocarcinoma that is often fatal. The biology of the respiratory system leads to the bottom line for everyone: If you smoke, quit; if you don't smoke, don't start. Healthcare workers carry a special burden to proselytize by both word and example. In support of this mission, a growing number of major medical centers have banned smoking anywhere on their campus, indoors or out.

Instructional Goals/Learning Outcomes

1. Describe the primary functions of the respiratory system, and explain how the delicate respiratory exchange surfaces are protected from pathogens, debris, and other hazards.
2. Identify the organs of the upper respiratory system, and describe their functions.
3. Describe the structure of the larynx, and discuss its roles in normal breathing and in the production of sound.
4. Discuss the structure of the extrapulmonary airways.
5. Describe the superficial anatomy of the lungs, the structure of a pulmonary lobule, and the functional anatomy of alveoli.
6. Define and compare the processes of external respiration and internal respiration.
7. Summarize the physical principles governing the movement of air into the lungs, and describe the origins and actions of the muscles responsible for respiratory movements.

8. Summarize the physical principles governing the diffusion of gases into and out of the blood and body tissues.
9. Describe the structure and function of hemoglobin, and the transport of oxygen and carbon dioxide in the blood.
10. List the factors that influence respiration rate, and discuss reflex respiratory activity and the brain centers involved in the control of respiration.
11. Describe age-related changes in the respiratory system.
12. Give examples of interactions between the respiratory system and each of the other organ systems.

Teaching Strategies

1. Lecture and Lab Ideas

a. To connect the conditioning of inhaled air with the anatomy of the nasal cavity, use Figure 23–3(b), a coronal section through the airways of a cadaver. Show that the conchae provide substantial surface for warming, humidifying, and cleaning the inhaled air stream. Tell the students that the curved surfaces cause the air stream to spin, like a centrifuge, sending particulate matter outward to the walls where they stick to the mucus that coats the respiratory epithelium. The mucociliary escalator delivers the mucus with its load of embedded debris to the oropharynx. There, a "spit or swallow" action eliminates it.

b. Use Figures 23–4(a) and 23–4(b) to show that the tracheal cartilages only partly encircle the trachea. Ask the students to guess at the functional significance. Then, show them Figure 23–6(b) to show the esophagus closely applied to the posterior of the trachea. Make the point that as food passes down the esophagus, the absence of cartilage there facilitates passage of the bolus.

c. When considering the autonomic control of bronchiolar smooth muscle, point out that during "rest and repose" when the PSNS dominates, a narrowing of the airways is appropriate. During sympathetic activation ("fight or flight"), bronchodilation facilitates rapid air flow. Note that exercise triggers bronchodilation and also vasodilation in skeletal muscle.

d. When defining respiratory minute volume with the equation $\dot{V}_E = f \times V_T$, point out that its form parallels the equation for cardiac output, that is, a volume of fluid per minute equals volume/cycle $\times$ cycles/minute.

e. Henry's law connects concentration of dissolved gas to the partial pressure. Thus, a partial pressure difference (gradient) equates to a concentration difference. Thus, diffusion down a partial pressure gradient during external or internal respiration can be viewed as diffusion down a concentration gradient, a concept familiar to students who may be uncomfortable with the concept of partial pressures.

f. When describing the effects of CO_2, H^+, and BPG on oxygen uptake and release from hemoglobin, focus attention on the saturation curves at low P_{O_2}, where the effects of these modifying agents are most dramatic. Show how the quantity of bound oxygen at any given P_{O_2} is affected by these ligands.

g. Carbon dioxide is about 10 times more soluble in water than is oxygen largely because it chemically reacts with water to form carbonic acid, which dissociates to yield the highly soluble bicarbonate ion. Thus, most carbon dioxide in the circulation is carried as the bicarbonate ion, but the interconversion between CO_2 and bicarbonate is effectively instantaneous because of the astonishing turnover rate (400,000 sec^{-1}) of the enzyme carbonic anhydrase present within RBCs.

2. Analogies

a. The mucus layer of the respiratory epithelium is like sticky flypaper. Both trap potentially troubling particles (flies or microbes) from the surrounding air stream.

b. Consider the pharynx to be a large cavern, extending from the nasopharynx to the laryngopharynx. If you were hanging from the uvula, you could jump down onto the tongue and walk down into the larynx or anteriorally into the mouth, or climb superiorly into the nasal cavity. As you headed up toward the internal nares, you could even take a left or right turn and squirm up to the middle ear through the auditory tubes. If you went straight into the nose, you could work your way between the conchae and into the paranasal sinuses.

c. As you show the SEM of lung parenchyma in Figure 23–10(b), compare the alveoli to the bubbles in soapy foam. The contact areas between adjacent bubbles are flattened, like the septae between air spaces. This makes the lung an open, spongy tissue unlike any other.

d. Surface tension is like a thin membrane of rubber. Adding surfactant is like punching holes in the rubber membrane.

e. Saturation of hemoglobin is like filling a miniature four-well Chinese checkerboard. When all four wells contain marbles (oxygen molecules), the board is full (total saturation of oxygen-binding sites). Then, have the students imagine billions of these "boards" (individual hemoglobin molecules) to get the smooth curves, rather than a rectilinear four-step graph of percent of saturation vs. number of marbles.

3. Demonstrations

a. Have the students sit quietly with eyes closed doing nothing but paying attention to their breathing. Encourage a state of eupnea by suggesting quiet contemplation. No writing, talking, reading, etc. Ask, is exhalation passive (i.e., requires no muscular effort)? Does inhalation use the intercostal muscles or just the diaphragm? Palpate yourself to find out.

b. Have students grip their larynx and then swallow. Ask, what happens first? Lead them to recognize that initially the larynx moves up. Tell them, using Figure 23–3(c), that this action flips the epiglottis downwards to occlude the glottis, and thus protects the lower airways from aspirated solids or liquids.

c. To demonstrate the contribution that the nasal cavity makes to the resonant quality of speech, pinch your nostrils together and say something to that effect. The students get a laugh, and some will remember.

d. Demonstrate the effect of surfactants at lowering surface tension with a bowl of clean water with pepper sprinkled on the surface. If you touch the edge with surfactant (soap or detergent), the pepper pulls away as the surface tension is lowered by surfactant entering the surface. The retraction is a visible sign of surface tension at the air–water interface on the side away from the surfactant.

e. Another demonstration of surface tension is the capability of water striders, those insects that "walk" on water by harnessing the surface tension layer.

f. You can show the same effect with a small piece of aluminum foil rested carefully on the surface. After you add detergent (surfactant), it sinks to the bottom.

4. Applications

a. Point out that the lungs play a central role in the moment-to-moment regulation of pH. By raising or lowering the alveolar ventilation rate, the CNS, responding to chemoreceptors, can alter P_{CO_2} and thus change H^+ concentration. Write the association–dissociation reaction on the board so they see the simple and direct relationship between P_{CO_2} and H^+ concentration.

b. No doubt students have heard references made to upper and lower respiratory infections, but probably don't know where one stops and the other begins. Figure 23–1 makes the distinction explicit.

c. One way to discuss the hemoglobin–oxygen saturation curve is in terms of chemoregulation. Since the percent saturation changes very little between 60 and 100 mm Hg, it would offer no advantage for oxygen levels in this range to contribute to respiratory drive. On the other hand, the reflex control of ventilation is strongly activated by lower P_{O2}'s as the curve suggests is appropriate. This is why at higher oxygen levels, respiratory drive is almost entirely sensitive to P_{CO_2}.

d. "Cricoid pressure" to occlude the esophagus prevents aspiration of gastric fluids and solids. This maneuver, used routinely during endotracheal intubation, is effective because the cricoid cartilage completely encircles the larynx. Show Figure 23–4 to illustrate this.

e. The nicotine in cigarette smoke paralyzes respiratory cilia temporarily. During sleep, they recover motility and move the mucus sheet toward the oropharynx. When a smoker rouses, a reflex of coughing to clear the accumulated secretion is activated.

f. Long-term smoking leads to the shedding of cilia and the arrest of the mucociliary escalator. This condition is characteristic of chronic bronchitis, the most common chronic obstructive pulmonary disease (COPD). The other COPD of significance is emphysema where alveoli coalesce and airways collapse. These reduce gas exchange, leading to chronic hypoxia and hypercapnia. Both conditions make life miserable.

5. Common Student Misconceptions/Problems

a. Students who have heard of *cellular respiration* in a cell biology class are often confused to hear the word *respiration* used sometimes to mean ventilation, or gas exchange in the lung, or gas exchange in the tissues. Best to anticipate the problem and spell out the multiple meanings of the word.

b. Clarify that the pharynx is a cavity rather than a structure. It connects the internal nose, middle ear, paranasal sinuses, respiratory tract, and digestive tract. Point out that when one gets a sore throat, it may become an otitis media or sinus infection.

c. Even though the word *larynx* sounds like the word *pharynx*, it is a structure and not a space, though it is hollow and the opening is called the glottis. The laryngeal prominence is what is commonly called the "Adam's apple."

d. The consistency of the lung is difficult for students to picture. Compare the texture of the lungs to a sponge. Compare the cut surface of a dried lung with a slice of bread, full of tiny holes. The SEM in Figure 23–10(b) shows this clearly.

e. Students often confuse terminal and respiratory bronchioles, since *terminal* implies that they should be at the end. Point out that terminal bronchioles mark the *end* of the conducting portion, whereas respiratory bronchioles represent the *beginning* of the respiratory portion, where gas exchange (respiration) takes place.

f. The intrapleural space is normally less than 1 nm thick and is filled with a watery fluid. Adhesion between the visceral and parietal pleurae is equivalent to the small negative pressure found here. If the pressure there becomes atmospheric, the lung collapses, and this space expands dramatically.

g. When explaining the gas laws, begin with a qualitative presentation to help the students develop an intuitive grasp of the physics. Start with pressure. Encourage the students to create a mental image of gas molecules banging around inside a container; most will recognize that pressure results from the collisions against the wall pushing outward. Then, it becomes easier to grasp that if you cut off part of the volume but keep the same number of particles, the rate of hits will rise, and so the pressure. Now an expression like PV = constant will make better sense.

6. Vocabulary Aids

a. Many students mispronounce pharynx and larynx by inserting an "n" before the "y." The result is "fair ninks" instead of "fair inks."

b. The thyroid cartilage with its prow-like shape takes its name from a Greek battle shield called the *teiros* with a similar shape.

c. The word *cricoid* means "ring-shaped." It is the cartilage that makes a complete ring around the larynx. Its Greek root also gives us the word *circle* (helps to remember that the cricoid encircles the larynx).

d. A doctor using a tongue depressor to examine your throat and telling you to say "AHHH" is an example of *phonation*. There is no word produced, just a sound. On the other hand, if you mouth a word, as if you were lip-synching, then you are articulating. Speech requires both phonation and articulation.

e. Tidal volume flows and ebbs, like the ocean tides.

f. Eupnea (quiet unlabored breathing) literally means "breathing well."

g. The *turbin*ates (conchae) are not only twisted like a *turb*an, but they cause the air stream to spin like a *turbine*.

CHAPTER 24

The Digestive System

Introduction

Well over half of the people who seek medical attention have gastrointestinal complaints. This striking fact is a tribute to the complexity, diversity, and conspicuous impact of digestive system function and dysfunction. The clinical problems highlight many of the key structures and processes found along the digestive tube. The oral cavity may be affected, from tooth and gum problems to ulcers of the buccal and lingual mucosae. The pharynx may be inflamed ("sore throat"). Swallowing may be disturbed (*dysphagia*), for example, after a stroke. The lower esophageal sphincter may be too tight (achalasia) or too loose (gastric reflux). Erosive ulceration of the gastric or duodenal mucosa may lead to fatal blood loss if the submucosa becomes involved. Gallbladder disease may cause pain that mimics a myocardial infarction. *Ileus* may paralyze transport in the small bowel, and diabetic neuropathy may interfere with the large bowel. Varicosities of veins that drain the pelvic floor may intrude into the anal canal as hemorrhoids. Constipation or diarrhea may follow from osmotic and secretory disturbances as well as transport problems originating in the myenteric plexus and muscularis externa. Beginning your presentation of this topic with a litany of digestive miseries such as this should arouse interest in the topic. Finally, the textbook provides many clinical notes on digestion that can be used to connect the basic science of the digestive system to clinical practice.

Instructional Goals/Learning Outcomes

1. Identify the organs of the digestive system, list their major functions, describe the functional histology of the digestive tract, and outline the mechanisms that regulate digestion.
2. Discuss the anatomy of the oral cavity, and list the functions of its major structures and regions.
3. Describe the structure and functions of the pharynx.
4. Describe the structure and functions of the esophagus.
5. Describe the anatomy of the stomach, including its histological features, and discuss its roles in digestion and absorption.
6. Describe the anatomical and histological characteristics of the small intestine, explain the functions and regulation of intestinal secretions, and describe the structure, functions, and regulation of the accessory digestive organs.

7. Describe the gross and histological structure of the large intestine, including its regional specializations and role in nutrient absorption.
8. List the nutrients required by the body, describe the chemical events responsible for the digestion of organic nutrients, and describe the mechanisms involved in the absorption of organic and inorganic nutrients.
9. Summarize the effects of aging on the digestive system.
10. Give examples of interactions between the digestive system and each of the other organ systems.

Teaching Strategies

1. Lecture and Lab Ideas

a. To provide a unifying theme for the anatomy of the digestive system, establish that it is based on a tube whose wall follows a *standard histological plan*. It comprises the same four layers from the esophagus to the rectum. At each junction between major regions of the digestive system there is a *sphincter* that controls passage along the tube. To complete the picture, accessory organs that lie outside the tube secrete essential fluids into the tube. Figure 24–3 illustrates the tissue organization in a composite view for the small intestine and Figure 24–10(a) shows the layers clearly in the esophagus. Here, of course, the mucosa is covered by stratified squamous epithelium.

b. When discussing the functional significance of the stratified squamous epithelium in the buccal cavity and esophagus, point out that the strong interconnections, abundant keratin, and multiple cell layers in a stratified squamous epithelium equip it to function in a highly abrasive environment. In contrast, a simple epithelium is optimal for secretion and absorption. A columnar shape reflects the large cell volume needed to accommodate the cellular machinery (RER, Golgi, mitochondria, secretory vesicles, etc.) that those functions require.

c. Clarify that the gallbladder does not produce bile, but stores, concentrates, and releases the bile that is produced by liver cells. After cholecystectomy, bile is still produced but is no longer concentrated. This dilute bile emulsifies fat poorly, so patients often experience intolerance to dietary fat after its removal.

d. Before showing the realistic anatomy of Figure 24–1, introduce the GI tract as a linear schematic written on the board from left to right, using arrows to show the direction of transport:

   Oral Cavity → Pharynx → Esophagus → Stomach → Small Intestine → Large Intestine → Anus

   Extend the diagram by adding the *accessory organs* and *sphincters*. If you want to assign a group project, this diagram makes a good basis: Divide the class into groups and divide the GI tract among them. Require an integrated presentation by the group but with an identifiable role by each student. The goal is to give an account of the structure and function of each component as defined by how it contributes to digestion of a meal that includes starch, fat, and protein. Motility, sphincter action, secretion, absorption, and neural and hormonal regulation can be addressed.

2. Analogies

a. The hormonal and neural responses of the digestive tract orchestrate the movement and digestion of the chyme as it travels from one organ to the next. After the cephalic phase is over, the responses thereafter are governed by the volume and composition of the luminal contents. This is like the chyme traveling along an automatic car wash. Its presence sets off an automatic series of events (vacuuming, soaking, soaping, scrubbing, rinsing, drying, polishing). As it reaches each station, it turns on the new process and turns off the one it just left.

b. The epithelium of the buccal cavity, esophagus, and anus seems like it resulted from the skin being poked inwards at both ends during embryogenesis.

c. A liver lobule is like a walled city built on waterways (refer to Figure 24–20(b)). At each corner the portal triad stands like a way station sending blood in and carrying bile out. Blood flows sluggishly past the row houses (hepatocytes) plastered with stucco (sinusoid endothelium), a mixture of rich venous blood and fresh arterial blood. The houses grab up what they want including the remnants of red cells such as heme and iron. They put out numerous proteins (enzymes, hormones, transporters) into the waterway and drip bile out the alleyway at the rear through pipes (bile canaliculi).

d. The sphincters along the digestive tube are like security gates along a prison corridor. At each gate, the proper signal must be given for the gate to open, and for the prisoner (the chyme) to pass.

e. The gut is like an earthworm. Outside, a thin pellicle (serosa); then a thick muscularis that performs peristalsis; finally, a mucosal tube to break down and absorb what has been ingested.

f. Cirrhosis (hardening) of the liver occurs when hepatocytes die faster than they can be replaced. The collagen that replaces the cells is like the callus in a tree that has been hit by lightning. All the damaged tissue is filled in, but the loss of cellular function is irreversible.

3. Demonstrations

a. Tell the students that the human digestive system contains more neurons than the human spinal cord. Then tell them that the "gut has a mind of its own." That should reinforce the significance of the intrinsic nervous system for regulation. The CNS (via the ANS) establishes general activity level rather than providing moment-to-moment adjustments to motility and secretion. Short reflex loops (Figure 24–5) along with hormonal interactions between elements of the digestive tract (Figure 24–22) demonstrate these local influences.

b. At the beginning of lecture or lab have three Petri dishes with a layer of 1 percent plain gelatin on the lecture table (gelatin = denatured collagen derived from the dermis of pigs or cows). Knox's gelatin or equivalent is available at any grocery. Sprinkle two plates with equal amounts of meat tenderizer ("Adolph's" or the like) and leave the third untreated as a control. Tenderizers contain a broad-spectrum protease such as papain or bromelain. Have a panel of students confirm that both plates are equally dosed. With a spatula or scalpel mince the gelatin to imitate chewing (mastication) in one of the treated plates, and leave the other undisturbed. Promise the class that, at the end of the hour, they will see how digestion works (both mechanical and chemical). Invite the students up when your talk is over. The gelatin (protein) that got "chewed" will be largely liquified (chyme), but the other plate will only have started to digest. This result dramatizes that mechanical digestion boosts the effectiveness of chemical digestion. Finally, make the point that this chemical digestion with a store-bought protease occurs *outside* the body, just like the digestive system! Only after epithelial uptake do the amino acids released from dietary proteins become available *inside* the body to support metabolism.

c. Repeat the demonstration of the histology of the primitive gut tube as outlined in Chapter 4, Section 1a of this Manual.

4. Applications

a. Ask the class if anyone recalls which cranial nerve provides most of the parasympathetic control of the gastrointestinal secretion and motility (N X, the right vagus). Show Figure 24–12(a) and point to the huge nerve trunk that follows the esophagus into the abdominal cavity. It should be clear why early anatomists named this nerve "the wanderer."

b. The lingual frenulum helps hold the tongue in place. Rarely the frenulum may be too short or extend too far anteriorly in some children. Known medically as *ankyloglossia*, it is sometimes referred to as "tongue tie." It often interferes with swallowing or speech. The condition can be corrected surgically by snipping part of the frenulum to loosen its restriction of the tongue, but some physicians prefer adopting a "wait-and-see" approach because the condition often resolves as a child grows.

c. The enzyme amylase in saliva initiates starch digestion in the mouth. Some years ago, baby food manufacturers were called to task because of the large amount of starch fillers put into jarred baby food in the place of other, more nourishing (and more expensive) ingredients. The starch thickens the product. If a parent fed the baby directly from the jar and then put the unfinished portion in the refrigerator, the next day the food would be completely liquified. By introducing the baby's salivary amylase, the enzyme was able to turn the starch to sugars.

d. About 1 in 500 newborns suffer from *projectile vomiting*, forceful ejection of the stomach contents. Pyloric stenosis, a narrowing of the pyloric sphincter, may be the cause and requires surgical correction if it persists.

e. Figure 24–21(b) shows that the common bile duct and pancreatic duct join at the duodenal ampulla. This anatomical arrangement should make it clear why obstruction of this passage by gallstones may lead to pancreatitis.

f. The horizontal section through the upper abdomen in Figure 24–19(a) offers a view seen in CT scans.

g. The large impact of digestion on water balance is pictured in Figure 24–27. This overview will be helpful again in Chapter 27.

5. Common Student Misconceptions/Problems

a. Students naturally are familiar with ingestion and defecation but need to have their understanding of the intervening processes and organs explicitly addressed. Using the simple system schematic suggested above (or Figure 24–1), can they define each organ's role in digestion?

b. Students usually think that food that has been swallowed is "inside" them. Make it clear that the gut lumen is topologically *outside* the body. Only after foodstuffs are hydrolyzed into their component simple subunits and then transported across the epithelium are they actually *inside* the body and accessible to cellular metabolism.

c. Make the nature of food explicit: Food consists of the cells and cellular products of plants and animals and so comprises all the biochemical classes. Hydrolytic enzymes specific to each class of molecular constituent reduce foods to simple organic molecules followed by transmembrane uptake and distribution by blood and lymph. Those components for which humans lack a specific hydrolase can't be assimilated and so are eliminated in the feces as *fiber*.

6. Vocabulary Aids

a. The name for the parotid gland translates as "near the ear" (par-otid). Figure 24–7(a) shows this clearly.

b. A mnemonic device for the four types of teeth from anterior to posterior [***I***ncisors, ***C***uspids (canines), ***B***icuspids (premolars), ***M***olars] is ICBM (Intercontinental Ballistic Missile).

c. Wisdom teeth were given their name because they erupt at about the age of 18 to 21 years, when we're presumed to be wiser.

d. The three phases of swallowing (buccal, pharyngeal, esophageal) are named for the location of the bolus. Figure 24–11 pictures these clearly.

e. The *cardia* of the stomach is named for its closeness to the heart.

f. The *fundus* (from Latin, meaning "bottom of," or the part opposite) provides mixing space because it rarely gets full (except during festive meals).

g. The effects of the gastric enzyme *rennin* on milk proteins can be seen if you witness an infant spit up milk; the milk appears curdled. Rennin, a protease produced by newborn mammals, denatures milk proteins, which coagulate. It derives from a Middle English term meaning to curdle. Cheese making begins with curdling of milk by rennin. Be sure to distinguish rennin (pronounced with a short E) from *renin* (pronounced with a long E), the renal peptidase important in blood pressure regulation.

h. The *jejunum* and *ileum* make up most of the small intestine. The *duodenum* occupies only the first twelve finger-widths (duodenum means "twelve" in Latin). Jejunum means "empty" because this gut segment is commonly found empty of contents in a postmortem examination. Ileum means "twisted," referring to the extreme twisting of this bowel segment.

CHAPTER 25

Metabolism and Energetics

Introduction

The "metabolic syndrome" represents a health crisis afflicting an estimated **45 million** persons in the United States and comparable numbers in other advanced industrial societies. The American Heart Association describes the metabolic syndrome as a cluster of risk factors that include:

- Central obesity (fat accumulation around and *within* the abdomen)
- Blood lipid disorders that promote atherosclerotic plaque—mainly high triglycerides and LDL cholesterol
- Hypertension (130/85 mm Hg or higher)
- Insulin resistance or glucose intolerance
- Prothrombotic state (e.g., elevated circulating levels of fibrinogen or plasminogen activator inhibitor)
- Proinflammatory state (e.g., elevated C-reactive protein in the blood)

The underlying causes of this syndrome are obesity, sedentary lifestyle, and genetic predisposition. A person can't change their genes but they can bring their calorie intake and expenditure into balance, that is, maintain a balance between *catabolism* and *anabolism*. Metabolic syndrome raises the risk of coronary heart disease, stroke, type II diabetes mellitus, and peripheral vascular disease. All these maladies relate to the buildup of atherosclerotic plaque. Accumulation of triglycerides within adipocytes, while strictly speaking an anabolic process, is nonetheless a potentially deadly metabolic strategy. The September 2007 issue of *Scientific American* focuses on these contentious issues of food, nutrition, and health.

Instructional Goals/Learning Outcomes

1. Define metabolism, and explain why cells must synthesize new organic components.
2. Describe the basic steps in glycolysis, the TCA cycle, and the electron transport system, and summarize the energy yields of glycolysis and cellular respiration.
3. Describe the pathways involved in lipid metabolism, and summarize the mechanisms of lipid transport and distribution.
4. Summarize the main processes of protein metabolism, and discuss the use of proteins as an energy source.
5. Differentiate between the absorptive and postabsorptive metabolic states, and summarize the characteristics of each.

6. Explain what constitutes a balanced diet and why such a diet is important.
7. Define metabolic rate, discuss the factors involved in determining an individual's BMR, and discuss the homeostatic mechanisms that maintain a constant body temperature.

Teaching Strategies

1. Lecture and Lab Ideas

a. When approaching the study of any complex and unfamiliar topic such as metabolism and energetics, a student benefits from starting with the big picture. The first two figures in the chapter suit this purpose well. Figure 25–1 establishes metabolism as the sum of catabolism and anabolism. It links to the digestive system (Chapter 24), which makes small molecules available either as sources of energy (ATP) or as biosynthetic building blocks. The ATP powers synthesis and a host of cellular reactions that sustain the individual. Figure 25–2 provides a scheme for organizing cellular energetics before considering complicated metabolic cycles and pathways. Figure 25–2 is a "Navigator" for the flow of substrate through cellular energy production and shows the evolution of carbon dioxide and water, but oddly omits oxygen as the electron acceptor. After presenting the details of intermediary metabolism contained in Figures 25–3 to 25–11, use Figure 25–12 to provide a coherent summary of cellular metabolism and energetics.

b. Write the overall reaction of "carbohydrate" metabolism on the board:

$$\text{"}CH_2O\text{"} + O_2 \rightarrow CO_2 + H_2O + \Delta$$

"carbohydrate" + oxygen → carbon dioxide + water + energy

By including the last term "energy" in these equations (written as both the chemical symbol and word), this expression makes clear why cells would bother to convert carbohydrate to carbon dioxide and water. It is not these end products that drive the life processes but the chemical *free energy* made available by these reactions.

The free energy is captured in the terminal phosphoryl group of ATP. Glycolysis (anaerobic respiration) by itself liberates only a small part of the energy potentially available from catabolism of a glucose molecule (only 2 of 38 possible ATPs). This is the fundamental reason behind the powerful drive to breathe. Humans can't survive anaerobically!

Showing catabolism at this simple level helps a student understand that what is happening overall is the transfer of hydrogen atoms (a proton and an electron) from carbohydrate to oxygen, the electron acceptor. Thus, the hydrogen is oxidized and the oxygen reduced. In cells, this transfer of hydrogen occurs in multiple steps of successive reduction and oxidation involving an "electron transport chain" that relies on coenzymes (NAD^+ and FAD^+) derived from niacin and riboflavin as the actual carriers of charge. This is the link between vitamin intake and enhanced energy that advertisers of foods and supplements allude to.

c. Triglyceride molecules provide more energy (~9 kcal/g) than carbohydrates (~4 kcal/g) simply because fats contain many more hydrogen atoms per gram. In effect, carbohydrates are already partially oxidized. Carbohydrates already contain a mole of oxygen for each two moles of hydrogen, whereas the hydrocarbon chain of fatty acids has none.

d. Make the point that ~60 percent of the energy released from foodstuffs by catabolism is lost as heat, and thus is disordered energy unable to support anabolism. However, this heat elevates body temperature and so speeds up cellular activities. Warm-blooded creatures such as humans enjoy the important advantages of more powerful movements and more rapid neural processing compared to their cold-blooded competitors. Therefore, the "waste" energy is hardly wasted, in an evolutionary sense.

2. Analogies

a. The high-energy content of fats (which are mostly hydrocarbon) as compared to carbohydrates is like the high-energy content of ethane (C_2H_6) as compared to ethanol (C_2H_5OH). Ethanol, like a carbohydrate, is partially oxidized or "burned," and therefore contains less potential energy.

b. As an analogy to electron transport, suppose there is a large lake on the top of a mountain. The water there was lifted by sunlight just as photosynthesis raised the energy of carbon dioxide and water to form carbohydrate (and oxygen). Now imagine the water running down the hill to a series of dams with waterwheels attached. Thus, the potential energy at the top will be converted to useful energy in a series of small steps designed to extract the largest fraction of the potential energy. If there were only a single dam, the energy released would often be excessive and so be wasted.

c. The absorptive state is like harvest time when food is being gathered and stored. The postabsorptive state represents the lean times between harvests when stored provisions are used to sustain life. Insulin is the hormone of abundance, glucagon the hormone of fasting.

3. Demonstrations

a. A wet spirometer can be used to demonstrate catabolism directly. When the bell is charged with pure oxygen, and with the exhaled carbon dioxide removed by an absorbent cartridge, the rate at which the bell falls over time is a measure of oxygen uptake by the subject. Lying, sitting, and standing provide different basal states of oxygen consumption. Oxygen uptake during stationary exercise (treadmill, bicycle) can be compared to resting states.

b. That metabolism produces heat is easily demonstrated. Measure oral temperature accurately with a digital thermometer. Then have the subject run vigorously in place for a minute or two and remeasure oral temperature. Core temperature will have risen by a few tenths of a degree. This rise activates heat loss mechanisms conspicuously (increased skin blood flow, sweating).

4. Applications

a. The formation of lactic acid from pyruvate (see Figure 25–7) during heavy muscle activity when ATP consumption surpasses aerobic capacity creates an "oxygen debt" that will be repaid during recovery. That is one reason why ventilation remains elevated long after exercise ceases (driven by low blood pH due to H^+ from lactic acid). Be sure to call attention to this homeostatic mechanism after the exercise demonstration described above.

b. Phenylketonuria (PKU) is a genetic disease caused by an inability to properly convert the amino acid *phenylalanine* into tyrosine. PKU causes mental retardation, but this risk can be minimized if a child born with PKU is fed a diet that restricts their phenylalanine intake. Most states require that all newborns be tested for PKU ("wet diaper test"). Check out the label on the next diet soda you drink; you may see a warning addressed to people with PKU. The artificial sweetener aspartame (Nutrisweet) is a dipeptide of phenylalanine and aspartate and so must be avoided by those individuals.

c. People experiencing ketoacidosis due to poorly controlled diabetes mellitus may have breath that has a fruity smell. The odor is due to ketones that form as a result of cellular glucose starvation that forces cells to rely on catabolism of fatty acids. Severe ketoacidosis is life threatening. People on a "low-carb" weight-loss diet often have this odor, as do those who may be too ill to eat.

d. The "water of respiration" produced by aerobic metabolism plays a modest role in the water balance of humans but, for animals that live in a desert, this water is crucial.

e. The "food pyramid" concept for optimizing human nutrition (Figure 25–13; Table 25–2) has changed substantially over the years. Its structure represents an ongoing struggle and compromise between competing interests, pitting scientists of one stripe against another. Often, important economic interests try to influence the USDA's recommendations but points of view also reflect scientific uncertainty. It is difficult to establish what constitutes optimum human nourishment. Students seeking a research project might explore the evolution of the "food pyramid" as an example of the interaction between science and public policy.

f. Figure 25–9 offers a lovely overview of lipid transport and fates. The role of LDL ("bad" cholesterol) in delivering cholesterol to the periphery as well as the role of HDL ("good" cholesterol) in removing cholesterol and returning it to the liver is easy to see and discuss.

g. The Clinical Note on Fats and Cholesterol provides a compact essay on these very important and interesting issues.

h. Defects in the urea cycle (Figure 25–10(c)) due to mutant enzymes lead to elevated levels of ammonia in the body; this causes serious or fatal brain damage.

5. Common Student Misconceptions/Problems

a. Many students naively confuse minerals with vitamins. They may simply think of them as the "stuff" you get in vitamin pills. Clarify that minerals are *inorganic* elements such as iodine, phosphorus, zinc, calcium, and iron derived from the soil. We can't make elements but must get them from our diet. Point out that vitamins are required *organic* molecules that can be synthesized by some organisms but not by humans.

b. Students may not recognize that the biochemical purpose of digestion is to break down the macromolecules that constitute food into its subunits by hydrolysis and then to supply the body's cells with these small molecules. Some are then broken down (catabolism) to supply energy (as ATP) to support biosynthesis of new macromolecular cell components (anabolism).

c. The urea cycle (Cori cycle) as illustrated in Figure 25–10(c) is a multistep enzymatic pathway whose details are omitted in this figure. These details are beyond the scope of this course, but students should be told that the missing details are well understood. They will be grateful to have one less thing to learn!

d. Obesity represents a form of *malnutrition*, that is, an imbalance between nutrient intake and bodily needs. It is far more common today than protein-calorie malnutrition.

6. Vocabulary Aids

a. Pull the word *glycolysis* apart to show its meaning; *glyco-* denotes sugar while *-lysis* denotes dissolving.

b. *Metabolism* comes from the Greek, meaning "to change." The word derives from *meta*, meaning beyond, and *ballein*, to throw.

c. *Catabolism* comes from the Greek, meaning "to cast something down," as if to break it. In catabolism, cells break apart food molecules and capture the energy released when covalent bonds are split.

d. *Anabolism* begins with the Greek root *ana-* meaning upward, that is, to build up. In anabolism, cells build up molecular structures and raise up the energy content of molecules.

e. The vitamins, whose structures are now known, were named initially for letters of the alphabet at the time of their initial isolation from foods. Later, as their biochemical nature became known, they received chemical names (e.g., vitamin B_2 is now known as thiamine and vitamin C is called ascorbic acid).

f. *Ascorbic* literally means "against scurvy," since scurvy is the nutritional disease that develops from vitamin C deficiency. British sailors were nicknamed "limeys" because British ships were provisioned with limes to prevent scurvy. Citrus fruits are especially rich in vitamin C.

g. The name *triglyceride* points directly at the structure of fats: three fatty acyl groups esterified to the three OH groups of glycerol.

h. Gluco–neo–genesis is easy to remember and to understand. "Generates new glucose" is what it means, literally translated. Contrast this with glycogen–olysis, the lysis (dissolving) of glycogen.

CHAPTER 26

The Urinary System

Introduction

Most students begin this topic with the belief that the *primary* job of the kidney is to "purify" or "filter" the blood, that is, to remove toxic substances from the blood and excrete them into the urine. Although purification is one of the functions of the urinary system, it is NOT the most fundamental. The swollen appearance of kidney patients provides a powerful clue to what that function is: the maintenance of homeostatic conditions, that is, the defense of fluid volume and composition. Thus it makes good sense to make the control of body composition by the kidney the central theme around which to develop the topic. Thus, the nephron, the functional unit of kidney function, defines and defends the ionic, metabolic, osmotic, and volumetric properties of the blood. This group of homeostatic functions provides a unifying principle for understanding the role of the urinary system as you progress systematically along the nephron. The regulatory roles of the kidney can be extended by describing the kidney's other systemic actions: blood pressure regulation through the renin–angiotensin system, defense of body water and sodium stores, its role in stimulating erythropoiesis, and its contribution to calcium metabolism by controlling its loss in the urine and its uptake from the gut as regulated by the steroid hormone calcitriol.

Instructional Goals/Learning Outcomes

1. Identify the components of the urinary system, and describe the functions it performs.
2. Describe the location and structural features of the kidneys, identify major blood vessels associated with each kidney, trace the path of blood flow through a kidney, describe the structure of a nephron, and outline the processes involved in urine formation.
3. Discuss the major functions of each portion of the nephron and collecting system, and describe the major factors responsible for urine production.
4. Describe the factors that influence glomerular filtration pressure and the rate of filtrate formation.
5. Identify the types and functions of transport mechanisms found along each segment of the nephron, explain the role of countercurrent multiplication, describe hormonal influence on the volume and concentration of urine, and describe the characteristics of a normal urine sample.
6. Describe the structures and functions of the ureters, urinary bladder, and urethra, discuss the voluntary and involuntary regulation of urination, and describe the micturition reflex.

7. Describe the effects of aging on the urinary system.
8. Give examples of interactions between the urinary system and each of the other organ systems.

Teaching Strategies

1. Lecture and Lab Ideas

a. Clarify that the renal corpuscle comprises both the glomerulus and glomerular capsule. The glomerular capillaries represent the only place where an *arteriole* supplies blood to a capillary bed and an *arteriole* carries it away. This unique microcirculatory design optimizes filtration of plasma. Stress that anything that crosses from the blood to the capsular space has made it past three distinct barriers: the fenestrated endothelium of the glomerular capillary, the basal lamina, and the podocyte layer of the glomerular capsule. The filtrate (or more precisely, *ultra*filtrate, because the process separates plasma proteins from the smaller molecules and water) has been pushed from *inside* the body to the *outside* by the blood hydrostatic pressure. Make the point that whatever is filtered but not absorbed will be lost from the body in the urine. The only way to reclaim items from the filtrate is for the epithelial cells of the renal tubule to transport them, either passively or actively, back into the blood space. In other words, "recoup it or lose it."

b. The physical principles that govern filtration across the filtration barrier into the capsular space are the same as those that were used to account for lymph formation in an earlier chapter. The processes differ in important quantitative ways that include high capillary permeability due to fenestrations, high blood pressure, and a substantial backpressure from capsular pressure. Glomerular filtration is far more intense: about 25 percent of renal plasma flow is filtered!

c. Discuss the mechanism of the countercurrent multiplier by using Figures 26–13(b) and 26–13(c) to explain how it depends on certain properties of the nephron loop (loop of Henle):

   1. The two limbs of the loop parallel each other.
   2. The permeability of the loop varies along its length.
   3. The descending limb is permeable to water, and impermeable to solutes.
   4. The thick ascending limb is impermeable to water, but actively transports NaCl out of the tubule.
   5. This anatomical arrangement *multiplies* the limited power of the loop epithelium to create a solute gradient, and thus leads to a very substantial interstitial osmotic pressure. This high salt concentration is harnessed to "suck" water out of the collecting duct as it passes through the inner medulla on its way to the renal papilla.

   After laying down these basics, use Figure 26–15 to explain how ADH adjusts water loss in the urine by changing water permeability to maintain the osmotic pressure of plasma in the normal range.

d. Because the ureters enter the posterior of the bladder (where the wall is thickest) at a steep angle (to form the upper corners of the trigone), they take on a slit-like structure when they reach the mucosa. This opening into the bladder behaves like a sphincter. When the bladder contracts during micturition, the ureteral openings are pinched shut and so not much urine backs up into the ureters.

e. Summary Table 26–1 nicely couples cross-sectional views of the nephron regions along with concise summaries of their distinctive functions.

2. Analogies

a. The retroperitoneal location of the kidneys (plastered against the body wall behind the peritoneal membrane) is like a wall clock that got papered over in a room with new wallpaper.

b. Compare the size of a kidney to the size of a double deck of playing cards.

c. Filtration through the glomerular barrier is like filter brewing of coffee. Gravity (= blood hydrostatic pressure) is the driving force; a mixture of fluid and solids (= blood cells and plasma) is applied to the filter and the openings between the paper fibers (= glomerular filtration barrier). This analogy may be useful for students who have not taken chemistry and so may not have witnessed laboratory filtration.

d. The medulla looks like strands of combed hair owing to the parallel arrangement of the collecting and papillary ducts. The calyx is like a cup that catches the urine as it drips out of the renal papilla, like a cone from which melted ice cream drips.

e. The urinary bladder is like a one-chambered heart. The ureters are like venae cavae and the urethra is like the aorta. When the muscular wall contracts, the ureters are pinched closed like cuspid valves, and fluid is pressurized and expelled.

f. The renal artery and vein are like private freeways for exclusive use by the kidney.

g. The trigone is like a triangular funnel; fluid enters at the upper corners and flows down to the urethra at the bottom.

3. Demonstrations

a. To demonstrate the powerful effects of plasma osmolarity (i.e., hydration status) on ADH release, ask the students to recall the color and volume of their urine when they are very thirsty (and thus hypertonic) to what it looks like after they have drunk fluids in quantity. A small volume of dark-colored urine shows that ADH leads to recovery of water from the urine and consequently to a concentrated urine.

b. Another demonstration of ADH's ability to concentrate the urine is to consider the volume and color of urine that forms after alcohol consumption. Ethanol inhibits ADH release. The large volume and pale color of the urine that results show that ADH inhibition leads to water loss because the ADH-dependent concentrative mechanism is blunted. This is like a pharmacologically induced *diabetes insipidus.*

c. As you describe the medullary pyramids, emphasize that they are shaped like pyramids rather than like triangles as Figure 26–4(a) and Figure 26–4(b) might suggest. Remind the class that the sectioning process converts a 3-D object into a 2-D view that may be misleading.

4. Applications

a. Because the composition of urine has been determined by plasma filtration, selective tubular absorption and tubular secretion, metabolites of hormones, neurotransmitters, and other chemical markers accumulate in the urine. Analysis of urine can say a lot, not just about kidney function, but about the endocrine, nervous, reproductive, muscular, skeletal, and other systems as well. A familiar example of this application is testing for human chorionic gonadotropin (hCG) in urine as a presumptive test for pregnancy. Another application is forensic testing for the presence of metabolites of illicit substances.

b. Creatinine levels in the plasma are useful estimates of GFR, since the filtered load of creatinine is entirely lost in the urine. The normal creatinine plasma is 1 mg/dL, and so a reading of 2 mg/dL (twice as high) implies that GFR is *half* of normal. This makes measurement of plasma creatinine a valuable tool to keep track of GFR in kidney patients.

c. Loop diuretics (such as Lasix) work from the luminal surface by blocking chloride carriers. This interferes with salt movement in the nephron tubule, especially in the ascending limb of the nephron loop. This leads to increased salt and water loss.

d. A Foley catheter placed into the bladder through the urethra drains urine in patients otherwise unable to urinate. This prevents painful accumulation of urine that might back up the ureters, enter the renal pelvis, and by raising capsular hydrostatic pressure, to slow or stop glomerular filtration.

e. The urethra is much shorter in females than in males. This may be the reason why bacterial infections of the bladder are far more common in females than in males.

5. Common Student Misconceptions/Problems

a. A few naïve students may be confused about excretory functions "down there" and hold the mistaken belief that there is a link between the digestive and urinary systems. Clarify this issue by reference to Figure 26–1.

b. As you present the functional anatomy of the nephron tubule shown in Figure 26–6, stress that this is a simplified representation: The blood vessels have been omitted, and the tubule has been shortened. Use Figure 26–7(a) to show a more realistic range of nephron morphology, contrasting cortical with juxtamedullary nephrons. Finally, use Figure 26–7(b) and (c) to bring out the intimate and specialized relationship between the nephron and microcirculation.

c. Often after students learn that ADH increases water permeability of the collecting duct, this somehow makes them think ADH leads to a dilute urine. To ward this off, use Figure 26–15(a) and (b) to illustrate the mechanism by which ADH leads to recovery of water from the filtrate in the collecting ducts and thus leads to a hyperosmotic urine.

d. Make it clear that proximal and distal refer to how far along the renal tubule from the corpuscle these structures are. "First" and "second" would have been more pedagogically useful terms!

6. Vocabulary Aids

a. Two words naïve students easily confuse are secrete and excrete. Make it clear that it is tubule cells that *secrete* substances into the urine, whereas *excrete* refers to elimination of wastes from the body in the form of urine, sweat, and feces.

b. Another pair of words that can be confused is ureter and urethra. No nifty mnemonic here, just learn it!

c. *Transitional* epithelium, which lines the ureters, bladder, and urethra, undergoes a *transition* in cell height and length to accommodate changes in the volume of urine within.

d. The renal corpuscle ("little body") comprises both the *glomerulus* (from Latin for "little ball") and the glomerular (Bowman's) capsule (from Latin *capsula,* "little box").

e. Interstitial (literally, "standing in between") is a term useful now when talking about the production of hyperosmotic urine, but will be useful in the next chapter as well, when discussing the components of the extracellular fluid compartment.

f. This chapter adopts many new terms for kidney anatomy based on *Terminologica Anatomica*. Bowman's capsule becomes glomerular capsule; renal capsule becomes fibrous capsule; interlobular vessels become cortical radiate vessels; loop of Henle becomes nephron loop.

CHAPTER 27

Fluid, Electrolyte, and Acid–Base Balance

Introduction

These topics can serve as a fruitful way to pull together many of the elements of this course. Nearly every organ system impacts body fluid volume and composition and *vice versa*. Also, fluid, electrolyte, and acid–base controls are intrinsically interconnected. To get students to think systemically and concretely about fluid balance, present a situation that is personally familiar to them, such as extreme water loss from sweating heavily on a hot day. Get them to state the basic problem (reduced total body water) and any symptoms they can imagine (intense thirst, dizziness, rapid pulse, etc.). Ask the students to consider what might happen in that case; for example, does water shift between the inside and outside of cells? Indicate that you will come back to these issues at the end of the topic to see if they can apply their new "fluids" knowledge to puzzle out the body's orchestrated response to severe dehydration. The negative feedback loops shown in Figure 27–5 will help them organize a systemic, mechanistic approach to these issues.

Instructional Goals/Learning Outcomes

1. Explain what is meant by the terms fluid balance, electrolyte balance, and acid–base balance, and discuss their importance for homeostasis.
2. Compare the composition of intracellular and extracellular fluids, explain the basic concepts involved in the regulation of fluids and electrolytes, and identify the hormones that play important roles in fluid and electrolyte regulation.
3. Describe the movement of fluid within the ECF, between the ECF and the ICF, and between the ECF and the environment.
4. Discuss the mechanisms by which sodium, potassium, calcium, and chloride ion concentrations are regulated to maintain electrolyte balance.
5. Explain the buffering systems that balance the pH of the intracellular and extracellular fluids, and describe the compensatory mechanisms involved in the maintenance of acid–base balance.
6. Identify the most frequent disturbances of acid–base balance, and explain how the body responds when the pH of body fluids varies outside normal limits.
7. Describe the effects of aging on fluid, electrolyte, and acid–base balance.

Teaching Strategies

1. Lecture and Lab Ideas

a. Review Figure 27–2 systematically with the class. This chart puts in graphic form the striking differences in ion abundance in the extracellular and intracellular compartments. These asymmetries underlie much that is vital to body function.

   1. The concentration of sodium ion is much higher *outside* of the cell, while that of potassium ion is much higher *inside* the cell. Ask the students, what does this asymmetry imply about the plasma membrane? (A permeability barrier; Na–K pump that creates and sustains the difference.) Point out that plasma membranes are generally more permeable to potassium ion, and thus the size and direction of the potassium equilibrium potential dominates resting transmembrane potential, putting it at about –90 mV.
   2. For divalent cations, calcium ion dominates the ECF and magnesium ion dominates the ICF. Magnesium is a necessary cofactor in many intracellular enzymatic reactions. This contrasts with using small changes in cytoplasmic calcium ion concentration to control contraction, membrane permeability, exocytosis, apoptosis, and other cellular mechanisms.
   3. Among anions, chloride ion concentration is higher in the ECF because chloride distributes passively, being expelled from the cell by the negative potential inside (opposites attract, like repels). Phosphate ion concentrations are higher inside the cell. Most of this "phosphate" is part of metabolic intermediates (e.g., alpha-glycerophosphate) and macromolecules (nucleic acids).
   4. Another big difference is that much of the negative charge inside cells is due to negatively charged amino acid side-chains such as histidine on acidic proteins. The text mistakenly attributes protein charge to alpha-amino and alpha-carboxyl groups. These are unavailable in proteins because they are incorporated in peptide bonds. In addition they are equal in number and so can't contribute to net charge. Amino acids are neutral zwitterions at physiological pH. Such proteins are in high concentration inside the cell not only because they are synthesized and function there, but also because their large size keeps them there.

b. Remind the students that the effects of ADH, aldosterone, renin–angiotensin, ANP, and BNP were discussed when discussing hormones and kidney function. Now the students should revisit those actions within the broader context of fluid and electrolyte balance. Illustrate how neuroendocrine reflexes contribute to osmotic balance with the ADH release in response to a rise in plasma osmolarity.

c. As you present the reaction of water and carbon dioxide (as catalyzed by carbonic anhydrase) to form carbonic acid and then hydrogen ion and bicarbonate ion, stress the reversibility of each step. Show how an excess of carbon dioxide will drive the reaction toward the right whereas a fall in pH will drive it toward the left.

d. To understand shifts of water between the ECF and the ICF, state that the two compartments are separated by plasma membranes that contain abundant aquaporin channels but few (open) ion channels. That is, the membrane is "semipermeable." These properties (high water permeability, low solute permeability) lead to osmotic equilibrium between the two compartments. Any change in osmotic pressure in one compartment is quickly eliminated by water flux until osmotic equilibrium is restored.

e. Relying on Figure 27–11, make it explicit that acid–base compensation relies on only two organs, the lung and the kidney. The lungs adjust carbon dioxide whereas the kidney controls excretion of hydrogen ion and bicarbonate ion.

f. The text (Figure 27–1) adopts a more realistic estimate of the proportions of the ECF and ICF. Older models estimate the ECF:ICF ratio as 1:2, but here it is given as 44:56. The caption acknowledges the difficulty of estimating these values experimentally. Many clinical applications adopt the older model.

g. Figure 27–3 offers a clear graphical overview of the principal fluid compartments, their size, and the fluxes between them.

2. Analogies

a. The text describes the cells of the body as being like fish living in a pond. They are dependent on the physical and chemical conditions imposed by the pond.

b. Osmotic pressure operates like a vacuum; it behaves as a *negative pressure* that **sucks** water across the plasma membrane from a less-concentrated compartment.

c. Water movement between compartments, driven by differences in osmotic pressure, is like air movements between different compartments in an inflatable mattress. The total amount of fluid doesn't change; it simply moves from one compartment to another, driven by differences in pressure.

3. Demonstrations

a. Encourage the class members to take a deep breath and hold it as long as they can. The "air hunger" produced by the breath hold shows how rising carbon dioxide partial pressure powerfully stimulates ventilation. This response to hypercapnia reflects how the nervous system adjusts alveolar ventilation to maintain constant pH and pCO_2. Should a parent worry when a petulant child threatens to suffocate themselves by holding their breath?

b. Encourage the class to hyperventilate (rapid deep breathing) to blow off CO_2, which will make them alkalotic. Why do paresthesias (tingling sensation) occur?

c. The evolution of bubbles of CO_2 when acid is added to baking soda (sodium bicarbonate) demonstrates the reaction whereby bicarbonate "buffers" excess hydrogen ions by forming carbonic acid and then carbon dioxide. Write out the reaction on the board to show how a "volatile" buffer system works. A volatile buffer system includes a gas phase reactant.

4. Applications

a. To expand circulating blood volume (e.g., when patients have hemorrhaged) isotonic saline is infused intravenously. Sodium is largely confined to the ECF, and so most of saline infusion's effects are on blood volume. Slowly, the blood volume falls as some of the saline leaves to enter the interstitial space.

b. IV therapy is usually postponed until hemorrhage can be stopped. Injected saline or blood will increase blood loss by raising blood pressure.

c. Brain swelling after closed-head trauma causes a rise in intracranial pressure. This presses on cerebral arteries, reducing blood flow to the brain. Swelling can be acutely reduced by intravenous injection of a solute such as mannitol (Osmotrol), a slowly metabolized sugar alcohol. Prompt the class to puzzle out what properties such a solute should have (nontoxic, not rapidly metabolized, permeable to endothelium, impermeable to plasma membranes). The therapeutic goal is a 30 mOsm increase in plasma osmolarity. Physicians estimate the quantity to add to the ECF using calculations like those in this chapter. Mannitol also induces osmotic diuresis since it is filtered but not reabsorbed. Taken by mouth, mannitol acts as a mild osmotic laxative useful in children.

d. Cellular swelling occurs after injury, a fundamental pathological reaction. It is caused by (1) a reduction of ATP generation, leading to (2) a weakening of sodium pump action, leading to (3) intracellular sodium accumulation, leading to (4) water entry, leading to (5) cellular swelling.

e. In IV fluid therapy, isotonic dextrose (i.e., 5% glucose) is injected when rehydration with water alone is the therapeutic goal. The glucose prevents osmotic injury during injection that pure water produces but is quickly consumed by metabolism, leaving the solvent water behind.

5. Common Student Misconceptions/Problems

a. Remind students about solutions (solvent +solutes), and point out that concentration (amount of solute in a particular volume of solvent) can change either by gaining or losing solvent (water) or by changes in the amount of solute (e.g., eating salty food). Figure 27–4 illustrates that a more concentrated solution means the solute molecules are crowded closer together, and *vice versa*. The molecules will be closer together or further apart depending upon how many solute molecules are present OR how much solvent volume there is. This is partly because electrolyte balance and water balance are intimately linked in homeostatic disturbances and their compensation.

b. Expressing hydrogen ion concentration in pH units will be troublesome for many A&P students. They frequently lack the math or chemistry background. Start by presenting the units in chemical terms (e.g., 10^{-7} M or 1×10^{-7} M hydrogen ion concentration). Circle the exponent and then say this is called "pH 7." Now ask, if the pH were 7.4, what would the hydrogen ion concentration be in chemical units ($10^{-7.4}$ M or 6.1×10^{-8} M)? In other words if the pH value is higher, the hydrogen ion concentration is lower. Take the class through the example of pH 6.8 applying the same reasoning. Finally, state that pH 7 is neutral while pH 7.4 is slightly basic and pH 6.8 is slightly acidic.

c. Make clear that "acidosis" most often involves a pH in the alkaline range (> pH 7) but "acidic" *compared to* pH 7.4, the normal pH.

d. Water moves passively through aquaporin channels driven by osmotic forces. There is no active "water pump," only ion pumps.

6. Vocabulary Aids

a. The term *normal saline* seen on some IV bottles or irrigation fluids means an isotonic NaCl solution (about 0.15 M). "Normal" here means a solution whose osmotic pressure is comparable to the normal osmotic pressure of the ICF.

b. "Electrolyte" implies charge-carrying capability, that is, the movement of ions.

c. "Acid" comes from the same root as "axe," being something sharp. *Oxy*gen (literally "acid-forming") shares the same root.

d. “Alkali” comes from the Arabic word *al-qily* that means “fire ash” or “potash,” the alkaline residue of burned wood. The science of chemistry was greatly advanced by Arabic scholars.

e. The textbook repeats this oft-given origin for the term “pH”: “The “p” in pH comes from the German term *potenz*, meaning power. Hence, pH refers to the **p**ower of **H**ydrogen.” In fact, Sørensen, who introduced pH in 1916, used the term *potenz* to mean exponent (a power of 10). The German word for power—in the sense of strength—is not potenz but rather is *kraft*.

f. The chemical symbols for sodium and potassium derive from their Latin names, ***Na***trium and ***K***alium.

g. Sodium ion loss in the urine is called *natriuresis* and is produced by natriuretic peptides (ANP, BNP). The word is formed from the Latin word for sodium (*Natrium*) and the Greek word for urine (*ouron*).

CHAPTER 28

The Reproductive System

Introduction

A natural, if surprising, unifying theme for this topic is the similarity in many respects between the male and female reproductive systems. Most laypersons embrace the French saying *vive la difference,* but the anatomy, physiology, and embryology argue against this commonsense view. As you begin your presentation, alert the students that you will be showing them how much the two genders share anatomically and physiologically. Examples of these parallels include the molecular signals that control sexual function, gonadotropin-releasing hormone, follicle-stimulating hormone and luteinizing hormone, the basic scheme for gametogenesis, the homology between the glans penis and glans clitoris, the prepuce that covers the glans, the parallel between the scrotum and the labia majora, the paired corpora cavernosae and their crura, similarities between ovary and testis, and finally the strong parallels between male and female sexual arousal and orgasm such as contraction of the dartos muscles in the scrotum and penis or vagina, hyperemia of the external genitalia, and finally, rhythmic contraction of muscles of the pelvic floor, and the intense pleasure that accompanies these functions. Of course, gender-related differences are abundant and apparent such as the routing of the urethra, the subtle but powerful chemical differences in the sex steroids and their actions, and the ultimate difference, it is the mother that carries the developing fetus!

Instructional Goals/Learning Outcomes

1. List the basic components of the human reproductive system, and summarize the functions of each.
2. Describe the components of the male reproductive system and the roles played by the reproductive tract and accessory glands in producing spermatozoa; specify the composition of semen; and summarize the hormonal mechanisms that regulate male reproductive functions.
3. Describe the components of the female reproductive system and the ovarian roles in oogenesis; explain the complete ovarian and uterine cycles; outline the histology, anatomy, and functions of the vagina; and summarize all aspects of the female reproductive cycle.
4. Discuss the physiology of sexual intercourse in males and females.
5. Describe the reproductive system changes that occur with aging.
6. Give examples of interactions between the reproductive system and each of the other organ systems.

Teaching Strategies

1. Lecture and Lab Ideas

a. Most students are surprised to learn that the testes form high inside the abdominal cavity (Figure 28–2), and then migrate in a growth movement through the inguinal canal into the scrotum. The descent into the scrotum provides the somewhat cooler environment required for normal spermatogenesis. As the testes descend, they carry their blood vessels, lymphatics, nerves, and ducts—bundled within a sheath of cremaster muscle fibers as the spermatic cord—as well as an investment of peritoneal membrane that becomes the *tunica vaginalis.* Also, point out that the testes do not induce the scrotum to form. The scrotum develops by growth of a pair of skin folds (homologous to the labia) that then fuse on the midline. The *raphe* (see Figure 28–3) remains as a sign of how the fetal scrotum formed. In females these skin folds do not normally fuse; they become the *labia majora.*

b. Accurately depicting the three-dimensional organization of the ductus deferens, seminal glands, prostate glands, and bulbourethral glands with two-dimensional text figures is difficult (Figures 28–1, 28–3, and 28–10) but three-dimensional models make it easy. Provide 3-D models in lab for study.

c. Point out that the corpus spongiosum extends into the glans penis and so it too expands and becomes hard during erection. This assists penetration into the vagina.

d. The hypothalamus drives the ovarian cycle, and so try to clarify the complex interactions between ovarian hormones and GnRH and gonadotropin secretion. Alternate between Figures 28–25 and 28–26 to show the changes in ovarian function and the regulatory pathways. Make it clear that the dramatic change in the functional zone of the endometrium represents a response to estrogen and progesterone, and that progesterone withdrawal as the corpus luteum involutes due to the drop in LH brings on the menses.

e. To dramatize a significant difference between males and females, when detailing meiosis in females using Figure 28–15, point out that meiosis in oocytes begins in the fetus and is completed perhaps 50 years later in the uterine tube at the time of fertilization!

2. Analogies

a. The prostate is like a fist wrapped around the prostatic urethra. During arousal it squeezes the urethra shut cutting off the flow of urine.

b. The description of the uterus as "pear-shaped" is common, but it looks more like a lightbulb to some students. An upside-down pear is more apt, but this view is misleading as well, since in sagittal section (see Figure 28–13), the uterus is seen bent forward over the bladder, more like a mountain shrub bent by the prevailing winds.

c. The fundus of the uterus is like the fundus of the stomach: Both are the "humps" of a hollow organ positioned opposite the outlet.

d. Students are often curious about whether urine might pass through the male urethra during emission and ejaculation. Assure them that the contraction of the smooth muscle within the prostate acts to pinch off the urethra, somewhat like an additional sphincter. This function prevents urine with an uncertain chemical composition from mixing with sperm, which require a defined chemical environment to function normally.

3. Demonstrations

a. Point out that males and females are both born with primordial mammary tissue, but without estrogen stimulation, it fails to develop in males. Administered estrogen forces breast development in men. Estrogen treatment is part of male-to-female gender change, along with orchiectomy, penis amputation, and surgical construction of a functional vagina from the patient's own tissues.

b. The testis and ovary share several similarities. A connective tissue capsule, the *tunica albuginea,* surrounds both. Both rely on mitosis and meiosis to produce gametes. Both respond to FSH and LH. Both produce sex steroids. But differences abound as well. The testis produces spermatozoa almost without limit whereas the number of ova is fixed before birth. Also, although gametogenesis involves two meiotic divisions in both, each primary oocyte yields a single ovum rather than four spermatozoa.

c. Wrap two fingers around a finger to show how the prostate surrounds the urethra.

d. The two mechanisms of egg transport by the uterine tube are demonstrated in Figure 28–17. Panel b shows the layers of smooth muscle in the wall that produce peristalsis; panel c shows the carpet of cilia whose action sends the ovum, surrounded by the corona radiata, tumbling along the epithelium toward the isthmus.

e. To dramatize the importance of chemical structure in physiological regulation, when showing Figure 28–24 point out that the only difference between testosterone and estradiol is the loss of a methyl group and the gain of a hydrogen atom. One produces manly traits, and the other womanly!

4. Applications

a. The cremaster reflex that elevates the testis in a thermoregulatory reflex can often be triggered by stroking or scratching the inside of the thigh. This reflex is usually absent in cases of testicular torsion and so is useful in diagnosing this painful medical emergency. Torsion of the spermatic cord leads to testicular death due to ischemia.

b. During a digital rectal examination, the posterior two-thirds of the prostate gland is palpated through the wall of the rectum. When showing Figure 28–1 point to the spatial relationship between the prostate gland and the rectum. Fortunately most prostate cancers develop in this region. Ultrasound images of the prostate also exploit this anatomy using a rectal imaging probe, as does surgery on the prostate.

c. Indicate that the fimbriae drape over the surface of the ovary, but the two are not physically connected. This means that an ovum may not be captured by the fimbriae and may become fertilized within the peritoneal cavity where implantation often occurs! This lack of physical continuity also means that pathogens can move from the vagina, through the cervix, along the uterine tubes, and enter the peritoneal cavity.

d. The normal acidic environment of the vagina protects against pathogens, but is incompatible with sperm survival. The bicarbonate ion buffer of semen promotes sperm survival by neutralizing hydrogen ions.

e. A substantial drop in sperm count and fertility results from long hot baths by men. The inhibition of spermatogenesis is unreliable as an effective contraceptive strategy but may interfere with fertility in couples who are experiencing conception difficulties. Also, boxer shorts are better than briefs at promoting spermatogenesis.

f. Sexual function is an example of sympathetic and parasympathetic dual innervation that provides cooperative control rather than the usual antagonistic control.

g. The menstrual periods can become so erratic during the time prior to menopause (perimenopause) that determining with certainty when ovulation has ceased is often impossible. For this reason, a woman is usually not considered to be menopausal until menstruation has stopped for one year.

h. Use Figure 28–24 to illustrate why aromatase inhibitors such as *tamoxifen* are useful for protecting against the reoccurrence of estrogen-dependent cancers of the breast and ovary.

5. Common Student Misconceptions/Problems

a. Students may believe that, like tubal ligation, vasectomy also requires abdominal surgery. Make it clear that the procedure is performed through a pair of small incisions in the scrotum. Figure 28–3 shows the spermatic cord in two states of dissection. Ask the class what might happen if the lymphatic vessel in the spermatic cord was mistakenly ligated and divided rather than the ductus deferens.

b. A longitudinal section through the penis can be misleading, because it gives the impression that there is only one corpus cavernosum. Make sure to clarify this by showing both longitudinal and cross-sectional views [Figure 28–11 (a) and (b)].

c. It sometimes confuses students to learn that FSH and LH also function in males. It may be helpful to point out that they have parallel functions in the two genders. FSH stimulates gamete development and LH stimulates steroid secretion (progesterone or testosterone).

d. Students may get the mistaken notion that an ovum develops from the primordial follicle to Graafian follicle in a single cycle. Clarify that the process takes several cycles to complete. That is why ova of varying states of development are found within the ovary at any time (Figure 28–16).

e. Most students think the menses mark the *end* of the menstrual cycle rather than its beginning. Point out that designating the beginning of a cyclic process is arbitrary. The onset of menses is taken as day 1 of the cycle because they are a readily observed marker of reproductive function.

f. Except for production of gametes, all cells divide by *mitosis. Meiosis* occurs only in gametes.

6. Vocabulary Aids

a. Clinically, the root "orchio-" for testis appears in such terms as *orchi*ectomy (surgical removal of the testis) and crypt*orchi*dism [an undescended ("hidden") testis]. *Orchis* is the Greek word for testis. Orchid plants are so named because their swollen roots resemble a testis.

b. Some students have difficulty distinguishing among the terms that include the root "sperm." Explain that spermato*genesis* refers to the entire process that leads to mature sperm. The process begins with the spermato*gonia* (*gonos* means "offspring" or "seed" in Greek). These are *stem cells* that lie at the base of the seminiferous tubules. Through the action of FSH on nurse or Sertoli cells, they undergo mitosis to form diploid spermato*cytes* (-cyte means cell) with 46 chromosomes. These then undergo meiosis to form haploid *sperma*tids with 23 chromosomes, which differentiate into spermato*zoa* (literally, "sperm animals") by the process of spermio*genesis* (literally, "giving birth to sperm"). Figure 28–8 pictures the steps in spermiogenesis.

c. Students associate the word *os* with "bone," and so may be puzzled to hear it used to indicate an opening in the cervical canal. Tell them *os* has two distinct usages in Latin. In this case, its meaning is "mouth" (as in "oral") and not bone (as in "osseous").

d. *Crus* means "leg" (of the *corpus cavernosum*).

e. "Haploid" derives from the Greek for "single." "Diploid" means "double."

f. "Estrogen" literally means "generates estrus." Estrus is called "heat" by laypersons; it is the time of sexual receptivity in lower mammals.

g. "Progesterone" literally means a ketosteroid (*-sterone*) that "favors" (*pro-*) gestation (*-gest-*).

h. *Vas*ectomy is named for the outmoded term for the ductus deferens (*vas* deferens), a misnomer based on the misidentification of the ductus as a blood vessel (vas).

i. An easy joke to make: Refer to the corpora *cavernosa* as the corpora *casanova.*

CHAPTER 29

Development and Inheritance

Introduction

From the instant of fertilization until the moment of death, sometimes ten decades later, coherent function never ceases, and each successive stage rests upon its predecessor. This chapter traces succinctly this series of orderly transitions, focusing in particular on the prenatal and perinatal processes. Some of what brings this about is understood, but much remains a profound mystery. Puzzling out the mechanisms by which the genome is selectively activated and inhibited to bring about the differentiation of hundreds of different cell types is one of the hottest fields in experimental biology. The forces and signals that drive partly differentiated cells from place to place within the embryo to ultimately form a human being plays largely as a baffling silent ballet. Our genetic inheritance drives the headlong march from birth to maturity and then to senescence and death. Exactly how poses a scientific problem that very likely will never be entirely solved. Don't be afraid to tell your students that our understanding of these long-term processes of development and maturation are poorly understood. Encourage your students to look beyond the intricate biology of individual development to seeing these processes in the context of the individual functioning within human society, and within a human world community. From neonate to doting grandparent, the individual's destiny is played out, limited and specified not only by their genes but also by the circumstances in which their genome operates. To appreciate the miraculous beauty of the development process, look at Lennart Nilsson's pioneering book of in utero endoscopic photography, *A Child Is Born* (Delta Publishers, 2004).

Instructional Goals/Learning Outcomes

1. Explain the relationship between differentiation and development, and specify the various stages of development.
2. Describe the process of fertilization, and explain how developmental processes are regulated.
3. List the three stages of prenatal development, and describe the major events of each.
4. Explain how the three germ layers participate in the formation of extraembryonic membranes, and discuss the importance of the placenta as an endocrine organ.
5. Describe the interplay between the maternal organ systems and the developing fetus, and discuss the structural and functional changes in the uterus during gestation.
6. List and discuss the events that occur during labor and delivery.
7. Identify the features and physiological changes of the postnatal stages of life.
8. Relate basic principles of genetics to the inheritance of human traits.

Teaching Strategies

1. Lecture and Lab Ideas

a. The cellular development and organogenesis, supported by maternal metabolism and physiology, that occur in the first trimester lay the basis for all the developmental steps that follow. If something disturbs these early developmental processes, then the effects of those disruptions will be apparent in every subsequent stage of development.

b. The chorionic villi intrude into the functional zone of the endometrium in the secretory phase, invading by means of lytic enzymes secreted by the trophoblast. The villi of the syncytiotrophoblast incorporate and envelop elements of the maternal circulation. Thus, the placenta acts as both "lung" and "gut" for the fetus. Umbilical vessels are named based on their relationship to the fetal heart and not the maternal heart. Thus, the fetus sends two umbilical *arteries* to the villi while a single umbilical *vein* returns oxygenated and nutrient-enriched blood to the fetus.

c. Alert the students to references within the text to the "Embryology Summaries" in Martini's *Atlas* that is bundled with the textbook. This material will be a big help in clarifying the relationship between fetal and adult structure. The origin and fate of the embryonic germ layers lie at the heart of organ function. Graphical explanation of gland formation and other derivatives of epithelia are alone worth the effort.

d. The second trimester provides for almost complete organogenesis (refer to Table 29–2 and the *Atlas*). Because of these many developmental events, by this time the fetus will have lost the undifferentiated look of the "generic" mammalian embryo, and become distinctly human in appearance (see Figure 29–8).

e. In the third trimester, the fetus "fleshes out," as organ systems become active; adipose cells load with triglyceride, and nerve and muscle cells provide for vigorous movement as any birth mother will attest. During this period, the fetus becomes more vulnerable to such hazards as loss of blood flow should the umbilical cord become kinked or twisted.

f. Review Table 29–1 with the students with two purposes: first, to show how the organs of an adult derive from the three germ layers, and second, to encourage the students to derive general principles about the fates of the germ layers. The "Embryology Summary" in the *Atlas* expands on this for each organ system.

g. Make it clear that implantation does not depend upon a maternal embrace by the endometrium ("the decidual reaction") but rather an enzymatic attack by the invading syncytiotrophoblast of the embryo. The cellular behaviors shown by the early embryo making a place for itself within the endometrium are not unlike a tumor. This ability to invade a suitable epithelium enables ectopic pregnancies, even within the peritoneal cavity.

h. The allantois, formed of endoderm and mesoderm, is the only extraembryonic membrane to persist in the newborn. It forms part of the urinary bladder. The *Atlas* summarizes organogenesis for each organ system.

2. Analogies

a. Gastrulation is like matter diving into a black hole.

b. The umbilical cord is like the air hose that connects an undersea diver with the "mother ship" overhead on the surface.

c. The chorionic villi dip into lakes of nutrient-rich maternal blood much like digestive villi extend into the nutrient-rich chyme in the lumen of the small intestine.

d. Human development is like the processes of biological succession as one moves from the graveled bank of a glacial stream where the simplest forms are growing to the simple aspen forest up the hillside to the complex mature forest that flanks it.

e. The umbilical cord is expanded and stiffened by deposits of hyaluronic acid much like the sponge rubber in the wall of the connection between an astronaut and his life support module. This substance is classified as *mucous connective tissue.*

3. Demonstrations

a. Introduce the Punnett square diagram to predict inheritance with an example. Expect them to solve genetic problems initially by mechanical manipulation, but as they work other examples, they should see a pattern. Using the Punnett square diagram, present the genetic problem. Start by saying that the point of the procedure is to predict the genetic characteristics of the offspring, given the genes of the parents. Analyze each alternative systematically to show the method.

   1. Write the phenotype and the genotype of the cross.
   2. Determine the possible alleles that each parent could contribute. This is the step that is frequently missed. Listing the alleles is a way of predicting all the potential genotypes. Point out that the analysis predicts the relative chance of a particular genotype, assuming random combinations.
   3. Develop the Punnett square and fill in the results. Point out that the pairs of alleles within the square represent the possible offspring genotypes.
   4. List the frequency of the genotype and phenotype. Explain that the frequency can be expressed as a percent or a ratio. Stress that the frequency is an *expected* outcome, independent of any *actual* outcome.

4. Applications

a. Pregnancy imposes a heavy burden on a woman's body as the various organ systems adapt to meet the needs of the pregnancy. These adaptations are why it is important for women to get prenatal care, so that their blood pressure, endocrine status, fluid and electrolyte balance, nutritional status, blood chemistry and morphology, and renal functions can be monitored and corrective actions taken as they become necessary. Refer the students to the textbook discussion of the impact of pregnancy on maternal organ systems.

b. Three-dimensional ultrasound provides very high spatial and temporal resolution images of the developing fetus in utero. Such images often help parents bond with their developing fetus, as well as provide information about developmental progress and identify any anomalies. Serious birth defects occur in about 1 in 100 cases.

c. Congenital defects can sometimes be repaired by surgical treatments conducted in utero. These techniques were first developed in operations on pregnant ewes at the same Scottish research center that was first to clone the sheep, Dolly.

d. Tissue obtained by chorionic villi sampling can be used to detect genetic, chromosomal, or biochemical anomalies. One cardinal advantage of chorionic villus sampling is that the results can be known many weeks earlier than by amniocentesis. Obtaining a CV sample poses a smaller risk to the fetus than amniocentesis and is less painful for the mother. It can be performed by inserting a plastic cannula through the vagina and cervical canal and then into the placenta, rather than by inserting a needle through the abdominal wall. Ultrasound control of either procedure makes them both relatively safe.

e. Slow intravenous injection of synthetic oxytocin (Pitocin®) can be used to induce or hasten parturition by stimulating myometrial contraction. Known as a "Pit drip" clinically, it substitutes for fetal oxytocin, the natural trigger.

f. Emphasize that the basic body plan is laid down very early in embryogenesis, well before a mother may be aware she is pregnant. Thus, it is important for the well-being of offspring that sexually active women of childbearing age adopt practices to protect them. These include adequate folate intake—to reduce risk of neural tube anomalies, modest alcohol consumption—to reduce risk of maxillofacial and neural anomalies, and avoidance of illicit drugs—to reduce risk of genetic injury and intrauterine death.

g. Measurement of pubis to fundus distance provides a useful measure of the progress of a pregnancy. This is illustrated in Figure 29–9b.

h. The first placental hormone to appear is human chorionic gonadotropin (hCG). This replaces maternal LH that declines and sustains the corpus luteum. Thus the embryo protects itself from being destroyed by the menses which would occur otherwise as the corpus luteum involutes and progesterone is lost. Detection of hCG in blood or urine is a common test for pregnancy. Other placental hormones include placental lactogen, relaxin, estrogens, and progesterone.

5. Common Student Misconceptions/Problems

a. The value of knowing about development and inheritance may not be obvious to the average A&P student. Thus, it is worth telling them explicitly that this knowledge will improve understanding of organ function in both health and disease. Support for this notion comes from the fact that tissues are categorized and the organs differentiate according to the germ layers laid down early in embryogenesis; the subdivisions and ventricles of the brain derive directly from the embryonic neural tube; tissue repair and organ regeneration rely on many of the same mechanisms seen during the growing process; the complexly folded mesenteries of the digestive tract are more easily understood if the formation of the primitive gut is learned first; malignant tumors are classified, diagnosed, and treated based on their embryonic cell line of origin; finally, point out that many diseases occur as a result of developmental problems.

b. Most students respond to the word *cleavage* with thoughts of a buxom woman. Share this thought while reviewing Figure 29–2. Clarify that the word refers to those early cell divisions where the maternal cytoplasm is divided among the "offspring." This stage remains essentially the same size overall and so each plump blastomere becomes progressively uniformly smaller with each succeeding division.

c. Implantation reflects the cellular and biochemical invasion of the endometrium by the blastocyst. The process is clarified in Figure 29–3. Students are often surprised to learn how long after ovulation this takes place.

d. Gastrulation is difficult for students to visualize. Figure 29–4 does a good job of "animating" the dynamic cell movements that create the primitive streak as ectodermal cells move toward the midline of the embryo and dive down to fill the middle layer of the early embryo with future mesoderm. A good independent learning project to encourage is an online hunt for useful video clips. A search at a site such as google.com using the strategy "gastrulation animation" will detect many useful resources.

6. Vocabulary Aids

a. The term *corona radiata* literally means, "radiating crown."

b. *Amphimixis,* the term for combining of the male and female pronuclei to complete fertilization, translates as "both mixed together." Amphimixis is the true "moment of conception."

c. "Morula," the bumpy ball of cells that forms early, means *mulberry* in Latin.

d. To distinguish genotype and phenotype, indicate that the "genotype" denotes the "type of gene," whereas "phenotype" denotes what can be seen (*pheno-* means, "to show"). Point out that phenotype results from an interaction between genotype and the environment within and without.

e. The word root *blast* occurs often in this chapter (trophoblast, blastomere, blastocyst), and so it is wise to tell the class that it comes from the Greek word *blastos,* which means "to bud." Thus, just as the bud turns into a leaf or flower, the blastocyst "buds" into an embryo.

f. "Syncytio-" literally means "same" (*syn*) "cell" (*cyte-*). In this region of the embryo in direct contact with the endometrium, the boundaries of the cells have been lost, and so the endometrium faces a sac of lytic cytoplasm.

APPENDIX

Changes in Terminology in the Eighth Edition

Chapter	Seventh Edition	Eighth Edition
Global	Apostrophe-s on eponyms	Removed apostrophe-s from most eponyms
1	medical anatomy	pathological anatomy
1	special physiology	organ physiology
1	protein fibers	protein filaments
1	exergonic	exergonic or exothermic
1	endergonic	endergonic or endothermic
2	dl (and throughout book)	dL
2	ml (and throughout book)	mL
2, 3, 12	cell membrane	plasma membrane
4	communicating junction	gap junction
4	button desmosomes	spot desmosomes
4	fibers	filaments
4	goblet cell	mucous cell
4, 25	lamina densa	dense layer
4	lipids	triacylglycerols
4, 8	fibrocartilage	fibrous cartilage
4	fibrocartilaginous pads	pads of fibrous cartilage
4	plasma cell	plasmocyte
5, 22	Langerhans cells	dendritic cells
5	liposuction	liposuction or lipoplasty
6	marrow cavity	medullary cavity
6	cancellous bone	cancellous or trabecular bone
6	fontanels	fontanelles (and throughout book)
8	ossa coxae	coxal bones (and throughout book)
8	os coxae	coxal bone
8	lunate bone, scaphoid bone, trapezoid bone, capitate bone, pisiform bone, and hamate bone	lunate, scaphoid, trapezoid, capitate, pisiform, and hamate
8	cuboid bone, navicular bone	cuboid, navicular
8	popliteal line	soleal line
8	Pott's	Pott
9	ellipsoidal joint	ellipsoid joint

Chapter	Seventh Edition	Eighth Edition
9	condyloid joint	condylar joint
9	metopic suture	frontal suture
9	humeroulnar joint	humero-ulnar joint
10	microfilaments	intermediate filaments
10	adrenal gland	suprarenal gland
10	regulated channels	gated channels
10	H zone	H band
10	resistance	load
11	lumbodorsal	thoracodorsal
12	lymphatic system	lymphoid system
12	lymphatic tissue	lymphoid tissue
12	chemically regulated	chemically gated
12	voltage regulated	voltage-gated
12	mechanically regulated	mechanically gated
14	intermediate mass	interthalamic adhesion
14	jugular ganglion	superior ganglion
14	nodose ganglion	inferior ganglion
14	mesencephalic aqueduct	aqueduct of midbrain
15, 18	adrenal medullae	suprarenal medullae
15	anterolateral	spinothalamic
17	external acoustic canal	external acoustic meatus
18	adrenal gland	suprarenal gland
18, 28	anterior pituitary gland	adenohypophysis
18	adrenal cortex	suprarenal cortex
18	chief cell, principal cell	parathyroid cell
18	posterior lobe	neurohypophysis
18	leptin and resistin	leptin (only); deleted resistin
19	pluripotent	multipotent
20	fibrous skeleton	cardiac skeleton
20	fast channels	fast sodium channels
20	troponin	troponin T, troponin I
21	metarteriole	precapillary arteriole
21	special circulation	vascular supply
21, 22	lymphatic system	lymphoid system
22, 23, 26, 28	hilus	hilum
22	subcapsular sinus	subcapsular space
22	Hassall corpuscle	thymic corpuscle
22	mobile macrophages	wandering macrophages
22	paratrabecular sinus	paratrabecular space
23	Type I cells	type I pneumocytes
23	Type II cells	type II pneumocytes
23	septal cells	type II pneumocytes
23	pleuropulmonary neoplasm	bronchopulmonary carcinoma
24	enterokinase	enteropeptidase
24	Brunner's glands	Brunner gland
24	mg/dl	mg/dL
24	upper dental arch	maxillary arch
24	lower dental arch	mandibular arch

Chapter	Seventh Edition	Eighth Edition
24	arch	arcade
25	C/g	Cal/g
25	C	Cal
25	kc	kcal
26	renal capsule	fibrous capsule
26	interlobular arteries	cortical radiate arteries
26	interlobular veins	cortical radiate veins
26	adipose capsule	perinephric fat (primary); perinephric fat capsule (secondary)
26	loop of Henle	nephron loop
26	Bowman's capsule	glomerular capsule
26	juxtaglomerular apparatus	juxtaglomerular complex
26	parietal epithelium	capsular epithelium
28	suspensory ligament	infundibulopelvic ligament
28	seminal vesicle	seminal gland (primary); seminal vesicle (secondary)
28	sustentacular cell	nurse cell
28	bulbourethral gland	bulbo-urethral gland
28	cervical os	external os of uterus
28	lateral (cardinal) ligament	cardinal ligament